P; 58,70,

Friedrich Kummer (Ed./Hrsg.)

Asthma

Structural Basis – Theophylline Today
Strukturelle Grundlagen – Theophyllin heute

Springer-Verlag Wien New York

Prof. Dr. med. Friedrich Kummer
2. Medizinische Abteilung mit Lungenkrankheiten und Tuberkulose,
Wilhelminenspital, Wien, Österreich

© 1995 Springer-Verlag/Wien
Printed in Austria

Druck: Ferdinand Berger & Söhne Gesellschaft m.b.H., A-3580 Horn
Printed on acid-free and chlorine-free bleached paper
Gedruckt auf säurefreiem, chlorfrei gebleichtem Papier – TCF

With 37 Figures / Mit 37 Abbildungen

ISBN 3-211-82670-X Springer-Verlag Wien New York

Preface

Asthma is a disease mainly based on eosinophilic inflammation with complex pathogenesis and a large spectrum of clinical manifestations.

Amongst a multitude of points of interest, recent research is directed towards the fascinating structural changes on the one side, and the new aspects of theophyllin on the other side. The 5th Vienna Asthma Forum 1994 was the latest in a series of biennial symposia, dedicated to ardent topics of bronchial asthma.

Structural changes of the bronchial epithelium fill most clinicians either with awe or with awkwardness. The speakers of our symposium, however, were able to fascinate the audience with hitherto unknown facts and images.

Dr. Konrad Morgenroth (Bochum) gave us a new insight into the early happenings within the cell after internalization of adeno-virus.

Dr. Len W. Poulter (London) found normals with hyperreactivity and no inflammation, shedding some light on cellular function how it is supposed to be.

The network of T-cell-regulation, macrophages and eosinophils was elucidated by the new data presented by Dr. Stephen J. Lane (London).

Dr. W. R. Pohl (Vienna) demonstrated the expression of heat shock protein from cultured epithelial cells.

Dr. Peter K. Jeffery (London) showed his unique ultrastructural visualization of activated cells within the epithelium.

Dr. Tari Haahtela (Helsinki) convinced the audience of the structural benefits that accompany clinical improvement during treatment with inhaled corticosteroids.

One of the most traditional compounds in asthma therapy, theophylline, was discussed in the second part of the symposium, which was opened by Dr. Jens D. Lundgren with a stimulating information on mucus secretion.

He was followed by Dr. Miles Weinberger (Iowa) who gave a remarkable surview of attitudes and rationale of theophylline in the past three decades.

Dr. John Costello (London) evoked the keen interest of the audience by his data on the late asthmatic reaction and activation of eosinophils, modulated by theophylline.

His talk was complemented by the latest results on cytokines presented by Dr. A. M. Vignola (Palermo).

Dr. Gilbert E. D'Alonzo (Philadelphia) was able to convey the message that theophylline has a particular impact on the chronobiology of asthmatic manifestations.

The speakers and the audience engaged in vivid discussions which could be reproduced and were included in full after each contribution.

Taken all together, the participants felt that this symposium was a unique opportunity to treat the subject of inflammatory modulation by theophylline in a very extensive and timely way.

I wish to thank our generous sponsors, Byk-Gulden and Mundipharma, represented mainly by Mr. Hubert Nettel und Mr. Ernst Novak, who engaged both in preparing the symposium and organizing this publication, which was realized with the dedicated help of Springer-Verlag and Mrs. Petra Naschenweng in particular. I wish to thank also Mag. Sujata Wagner, who did a marvellous job transforming the recordings into printable language. And finally, I thank my secretariat, Mrs. Eva Friedrich and Miss Natascha Kohlweiß, for so much patience and so many extra hours.

Vienna, January 1995   *Prof. Dr. Friedrich Kummer*

Contents/Inhaltsverzeichnis

Contributors/Autorenverzeichnis

Bellia, V., M.D., Istituto di Medicina Generale e Pneumologia, Università degli Studi di Palermo, Palermo, Italy

Bonsignore, G., M.D., Prof., Istituto di Medicina Generale e Pneumologia, Università degli Studi di Palermo, Palermo, Italy

Burke, C. M., M.D., James Connelly Memorial Hospital, Dublin, Irland

Chiappara, G., M.D., Istituto di Fisiopatologia Respiratoria, C.N.R. Palermo, Palermo, Italy

Costello, J., M.D., Prof., King's College, School of Medicine & Dentistry, Department of Thoracic Medicine, London, U.K.

D'Alonzo, G. E., D.O., *Professor of Medicine*, Temple University, Medical School, Philadelphia, U.S.A.

D'Amico, D., M.D., Istituto di Fisiopatologia Respiratoria, C.N.R. Palermo, Palermo, Italy

Haahtela, T., M.D., Department of Allergic Diseases, Helsinki University Central Hospital, Helsinki, Finland

Jefferey, P. K., MScPhD., Lung Pathology, Department of Histopathology, Royal Brompton National-Heart & Lung Institute, London, U.K.

Laitinen, A., M.D., Department of Pulmonary Medicine, University Central Hospital, Helsinki, Finland

Laitinen, L. A., M.D., Prof., Department of Pulmonary Medicine, University Central Hospital, Helsinki, Finland

Lane, St. J., M.D., PhD., Department of Allergy and Allied, Respiratory Disorders, Guy's Hospital, London, U.K.

Lee, T. H., M.D., Prof., Department of Allergy and Allied Respiratory Disorders, Guy's Hospital, London, U.K.

Lundgren, J. D., M.D. DMSc., Department of Infections Diseases (144), Hvidovre Hospital, University of Copenhagen, Hvidovre, Denmark

Merendino, A. M., M.D., Istituto di Medicina Generale e Pneumologia, Università degli Studi di Palermo, Palermo, Italy

Morgenroth, K., Prof. Dr., Ruhr-Universität Bochum Medizinische Fakultät, Abteilung für allgemeine und spezielle Pathologie, Bochum, Deutschland

Pace, E., M.D., Istituto di Fisiopatologia Respiratoria, C.N.R. Palermo, Palermo, Italy

Pohl, R., Dr. Univ.-Doz., II. Medizinische Abteilung, Wilhelminenspital, Wien, Österreich

Poulter, L. W., M.D., Department of Clinical Immunology, Royal Free Hospital, School of Medicine, London, U.K.

Romberger, D. J., M.D., University of Nebraska, Med. Ctr., Omaha, U.S.A.

Spatafora, M., M.D., Istituto di Medicina Generale e Pneumologia, Università degli Studi di Palermo, Palermo, Italy

Vignola, A. M., M.D., Università degli Studi di Palermo, Istituto di Medicina Generale e Pneumologia, Palermo, Italy

Weinberger, M., M.D., Professor of Pediatrics, Director, Pediatric Allergy & Pulmonary Division, Department of Pediatrics, University of Iowa Hospital and Clinics, Iowa City, U.S.A.

Electron-microscopic basis of bronchial hyperreactivity

K. Morgenroth

Department of Pathology, Ruhr University, Bochum, Germany

Summary

Ultrastructural changes are one of the bases of bronchial hyperreactivity. A number of experimental and clinical studies have shown that a nerval regulation is responsible for the adjustment of bronchial functions to changing environmental conditions. These nerval reactions function in a coordinated manner even in the denervated lung and in explanted bronchial mucosa, indicating that there is a special mechanism of regulation in the mucosa itself. If these autonomous and nerval regulatory mechanisms are disturbed, bronchial hyperreactivity may develop. The question arose as to whether a histomorphological substrate exists for this hyperreactivity and whether cellular changes are related to this dual system of regulation.

Systematic electronmicroscopic examinations were carried out in biopsies from children (aged 6 to 11 years) who were clinically asymptomatic at the time of biopsy. Based on our earlier clinical observation of an evident relationship between viral infection and the development of hyperreactivity, we induced viral infections in the explanted mucosa. Our intention was to observe early functional impairment following structural changes.

The biopsy specimens of these children with bronchial hyperreactivity consistently showed a sector-shaped intraepithelial edema.

The intercellular gaps had widened and the intercellular junctions were disconnected. These changes seem to be responsible for grave alterations in mucosal diffusion.

In these experiments with explanted mucosa, we observed the first contact of the virus with the cell membrane and its penetration into the cytoplasma. It was noted that the impact of the virus caused damage to the filaments and tubules of the cytoskeleton and simultaneous cessation of autonomous regulatory functions. The intercellular gaps became wider and the intercellular junctions were disrupted. Destruction of the cytoskeleton was accompanied by disorientation of the basal body of the cilia, which led to an uncoordinated ciliary beat.

The altered diffusion is manifested in changes in the transport of ions and increase of intraepithelial fluid, which in turn speed up the diffusion of inhaled allergens from the lumen into the subepithelial zones of connective tissue. Simultanously, there may also be direct irritation to the intraepithlial nerve endings. Disruption of intercellular junctions abolishes the horizontal spread of information within the mucosa and this, in turn, disturbs the coordination between the formation of secretion and ciliary transport.

These observations made from biopsies and experimental studies indicate that both components of the dual regulatory mechanism can be modified by viral infection so as to effect an increase in reactivity.

Zusammenfassung

Elektronenmikroskopische Grundlagen der bronchialen Hyperreagibilität. Die Reaktionsfähigkeit der Bronchialschleimhaut mit einer besonderen Anpassungsfähigkeit der Funktion an die sich ständig ändernden Umweltbedingungen unterliegt, wie dies eine Vielzahl von experimentellen und klinischen Befunden belegen, einer nervalen Regulation. Da die Funktionen jedoch auch an der denervierten Lunge und an der explantierten Bronchialschleimhaut koordiniert weiter ablaufen, muß ein eigenständiger, an die Schleimhaut gebundener Regulationsmechanismus bestehen. Diese autonomen und nervalen Regulationsmechanismen müssen gestört sein, wenn sich eine erhöhte Reaktionsbereitschaft im Sinne einer Hyperreagibilität entwickelt. Es stellt sich deshalb die Frage, ob ein histomorphologisches Substrat für die Hyperreagibilität existiert und in welcher Weise diesem dualen System der Regulation zelluläre Veränderungen zuzuordnen sind.

An Schleimhautbiopsien von Kindern mit hyperreagiblem Bronchialsystem ist im klinisch symptomfreien Intervall regelmäßig ein sektorförmig angeordnetes intraepitheliales Ödem zu ermitteln. Es besteht eine ausgeprägte Erweiterung der Interzellularspalten mit einer Lösung von Interzellularverbindungen. Die intraepithelial verlaufenden Nervenendigungen werden dabei freigelegt. Die Veränderungen sind wahrscheinlich die Grundlage für gravierende Änderungen der Diffusionsverhältnisse an der Schleimhaut.

Der mögliche Zusammenhang zwischen Virusinfektion und Hyperreagibilität wurde an explantierter Bronchialschleimhaut untersucht. Die Kontaktaufnahme der Viren mit der Zellmembran und ihr Eindringen in das Zytoplasma ist zu verfolgen. Als Folge des Virusbefalls der Zellen wird die filamentäre ünd tubuläre Struktur des Zytoskeletts zerstört, während gleichzeitig die autonom geregelten Schleimhautfunktionen eingestellt werden. Es entwickelt sich eine Erweiterung der Interzellularspalten mit Lösung der Interzellularverbindungen. Die Destruktion des Zytoskeletts geht mit einer Desorientierung der Basalkörper der Zilien einher. Die Koordination des Zilienschlages wird aufgehoben.

Die Änderung der Diffusionsverhältnisse mit einer Veränderung des Ionentransportes und eine Zunahme der intraepithelialen Flüssigkeit ermöglicht eine rasche Diffusion von Allergenen aus der Atemluft in die subepitheliale Bindegewebszone. Gleichzeitig ist eine direkte Irritation der intraepithelialen Nervenendigungen möglich. Durch die Lösung der intraepithelialen Zellverbindungen wird die horizontale Informationsausbreitung aufgehoben, die zu einer Störung der Koordination zwischen der Sekretbildung und dem Sekrettransport durch die Zilien führt. Die Beobachtungen am Biopsiematerial und die Ergebnisse der experimentellen Untersuchungen sprechen dafür, daß die beiden Komponenten des dualen Steuerungssystems an der Bronchialschleimhaut durch eine Virusinfektion so modifiziert werden können, daß eine übersteigerte Reaktionsbereitschaft im Sinne der Hyperreagibilität entstehen kann.

Introduction

Asthma bronchiale is a result of hypersensitivity of the bronchial system and is manifested in attacks of bronchial obstruction. The defense mechanisms of the mucosa are excessively exaggerated even after minimally irritative noxae. The various functions of the bronchial mucosa such as secretion, transport of secretion by ciliary movement, the tonus of bronchial musculature, their contraction during cough and their coordinated function, form one of the most significant and effective defense systems of the organism, especially for dealing with environmental changes. This defense system can adjust rapidly to environmental changes. The smooth coordination of the various components of this system and the balanced activity of its elements are responsible for the high efficiency with which the system functions.

This harmonious coordination requires a special kind of regulation. Hence, changes in reactivity that are manifested in the form of hyperreactivity must be caused by changes in the mechanism of regulation. There could be an imbalance between activity and aggressivity of the triggering agent on the one hand, and the reaction of the mucosa on the other. The question arises as to whether a substrate for this differentiated regulation can be derived by histomorphological methods. The second question is, which changes will bring about hyperreactivity.

A multitude of clinical, physiological and pharmacological observations indicate that the functions of the bronchial mucosa are controlled by nerval regulation. The fact that these functions are executed in a coordinated manner even in the denervated lung and in explanted bronchial mucosa indicates that there is an independent regulatory mechanism in the mucosa itself. It may be reasonably expected that both components of the dual regulatory system will be affected by hyperreactivity.

Experiments performed on animals and on the material derived from biopsies have proven the innervation of the various components of the bronchial mucosa (Morgenroth and Donner 1985, Andres and von Dürig 1985, Morgenroth 1992). Explanted human bronchial mucosa was examined systematically in various stages of functional alteration, by light- and electronmicroscopy, to ascertain the morphological basis of direct reception and spread of information in the mucosa.

Biopsy material from children (aged 4 to 11 years) having bronchial asthma and hyperreactive bronchial systems was available for a systematic evaluation of changes in hyperreactivity. The material was derived during clinically symptomfree intervals, from equivalent regions of the main bronchus.

Results

Light microscopy revealed intraepithelial edema in the biopsied material. The edema was located in sectors of different widths (Fig. 1). There was no inflammatory infiltration of the subepithelial connective tissue. Slight spreading of the basal membrane was found in circumscribed regions.

Electronmicroscopy revealed widening of the intercellular gaps from the basal membrane up to the apical section of the epithelium.

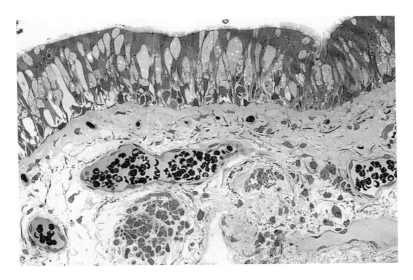

Fig. 1. Bronchial biopsy of a 4-year-old girl with bronchial asthma and a hyperreactive bronchial system. The biopsy was taken in a symptom-free interval. Sector-shaped distribution of intraepithelial edema. Otherwise the basic structure of the bronchial mucosa is intact. Semi-thin section. Staining: basic fuchsin and methylene blue. × 480

The zonulae occludentes remained closed in most cases. The cell connections were opened only in certain areas. The considerably widened intercellular spaces are interwoven by long and partly ramified intercellular bridges. Some of the intercellular contacts are disrupted and this causes the cell processes to float freely in the intercellular spaces (Fig. 2).

When the intercellular spaces are opened, the efferent and afferent nerve endings in the epithelium that lie immediately adjacent to the cell

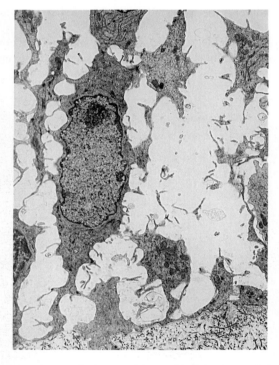

Fig. 2. Electronmicroscopic image of intraepithelial edema of the bronchial epithelium in a hyperreactive bronchial system. Same biopsy as in Fig. 1. Excessive widening of the intercellular gaps. Long and narrow cellular processes that float freely in the spaces or are connected to each other by their contact zones. Transmission electronmicroscopy. × 7.200

membrane are exposed, and become detached from the cell membrane. The degree of vacuolation of the epithelial cells varies. In addition, the basal bodies of the cilia are disoriented in some cells, and the surrounding basic filamentary and tubular structures of the cytoplasma are destroyed (Fig. 3).

The morphology of this region indicates grave changes in the diffusion of the epithelium. Irritants that are inhaled during respiration enter the sol phase of the layers without hindrance, and diffuse into the subepithelial zone of connective tissue. They reach directly to the nerve endings that lie in the intercellular spaces.

Clinical experience has shown that there is a close relation between viral infection of the respiratory tract and the development of hyperreactivity. Hence, the question arises as to which changes take place in the epithelial cells during the initial phase of the viral infection, and

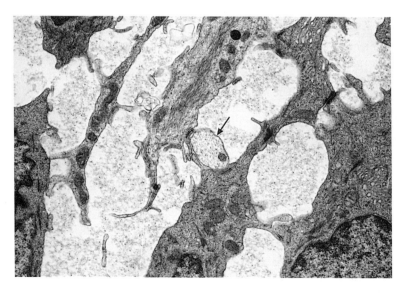

Fig. 3. Widening of intercellular spaces in the bronchial epithelium in a hyperreactive bronchial system. The increase of fluid causes the nerve endings in the epithelium to detach from the cell membrane of the epithelial cells. Thus they are exposed (arrow). Transmission electronmicroscopy. × 13.000

whether these changes correlate with those observed in the biopsied material.

Biopsy material is not available for examinations of this kind. Hence, systematic electronmicroscopic examinations were performed in cultures of explanted mucosa derived from surgical specimens. When evaluating the results, special attention was given to the behaviour of the cytoskeleton because the purpose was an exact evaluation of the significance of this system for autonomic regulation of the bronchial mucosa.

After viral infection, the epithelial functions that normally take place in a regular manner, were seriously altered for 12–24 hours. Ciliary beat was irregular at first and stopped completely afterwards. The regular secretory function of the epithelium was blocked.

In relation to these systematically registered functional changes, the cytoplasmic changes in the epithelial cells were also assessed. The alterations caused by the virus in the filamentary and tubular sytem of the cytoskeleton were registered, and their course was followed.

Marked vacuolation was noted especially in the apical sections of the cells. This was accompanied by reduction of the apical terminal filamentary system. In adenoviral infection, reduplication of the virus takes place in the nucleus. The viruses are transported within a vacuole system in the cytoplasma. The viruses in the cytoplasma are in different stages of maturity. The intercellular spaces open (Fig. 4) and the viruses are expelled from the cytoplasma into the intercellular spaces through the lateral part of the cell.

Reduplication of the virus takes place in the vacuole system during RSV infection as well. The precursors of the virus are transported to the filamentary system in the cytoplasma of epithelial cells. The virus precursors are revealed as a beaded structure in the filaments of the cytoskeleton. The destruction of filaments and tubules and their inadequate regeneration because of the metabolic changes that follow viral infection are probably responsible for the marked reduction of the cytoskeleton. The altered course and direction of the ciliary beat lead to disorientation of the basal bodies of the cilia in the cytoplasma (Fig. 5). The loss of tubular sections of the cytoskeleton is probably the reason for the fact that the basal bodies lose their anchorage in the cytoplasmic matrix.

The cytoskeleton plays an important role in the reception of information and in intra- and intercellular transfer of signals. This has been

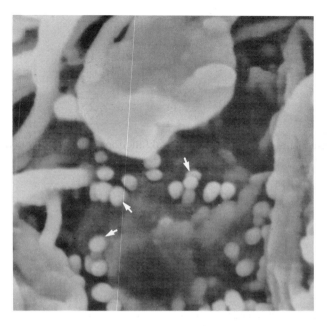

Fig. 4. Inital phase of viral infection in the bronchial epithelium. The viruses get into contact with the cell membrane of the epithelial cells (arrows) and freight the carriers of their genetic information into the cytoplasma of the cells. Grid electronmicroscopy. × 20.000

observed in several other organ systems as well. Hence, systematic visualization and reconstruction of this system in the bronchial epithelium were attempted. There is a highly sophisticated system of microvilli and microtubules in the epithelial cells of bronchial mucosa. The cytoplasmic matrix of the epithelial cells is permeated by microfilaments that are inserted in the external nuclear membrane. They are arranged in bundles in the cytoplasma and end in the contact zones of the cell membrane. This filamentary system is joined together horizontally over large surfaces, by macromolecular connections between the cell membranes.

There is a circular and extremely dense filamentary system in the apical sections of the epithelial cells. It could be called the marginal

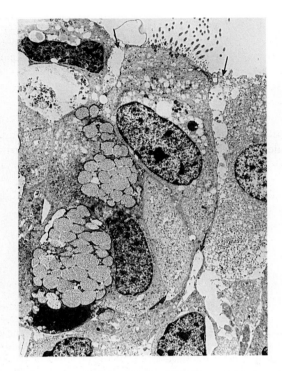

Fig. 5. Intracytoplasmic changes in the bronchial epithelial cells and reduplication of the virus. Vacuolation in the apical section of the epithelial cells. Widening of intercellular spaces (arrows). Precursors of the virus are detectable in the intracytoplasmic vacuoles. In adenoviral infection, the newly-formed viruses are expelled into the intercellular spaces. Transmission electronmicroscopy. × 4.800

apical filamentary system. In their horizontal course, the bundles of filaments take in other complexes of filaments that, in turn, are connected to the external nuclear membrane. The bundles of filaments join with the zonulae adhaerentes in the region of the apical cell membrane. Bunches of filaments and microtubules run from dense zones into the microvilli that are arranged on the cell surface, in a regular pattern between the cilia. The filaments and microtubules can be followed here up to the tips of the narrow cell processes.

Light- and electronmicroscopic examinations using immunohisto-chemical marking revealed that these filamentary structures consist of cytokeratins, whose spatial arrangement can be observed on a laser grid microscope (Phillipou et al. 1993) (Fig. 6).

In addition, there exists a system of microtubules arranged in a latticed fashion. These microtubules permeate the cytoplasma from the nuclear membrane up to the cell membrane and are characterized by irregular densities (satellites and centrioles). A ramification of the microtubules may be observed in these dense areas. They form a regular network around the pedical region of the cilia and are inserted in their lateral branches (Fig. 7). They end laterally in the dense inner zones of the cell membrane, in the region of the zonulae adhaerentes.

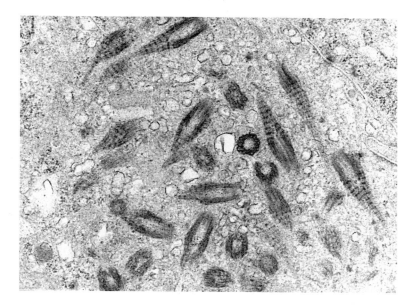

Fig. 6. Disorientation of the basal bodies of the cilia after viral infection. Loss of the tubules and filamentary system between the basal bodies. Transmission electronmicroscopy. × 39.000

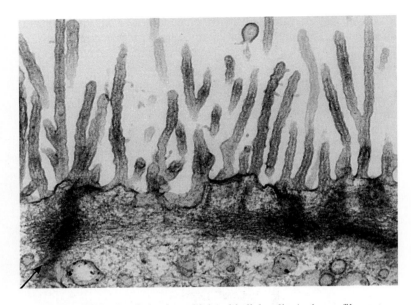

Fig. 7. Cytoskeleton of the bronchial epithelial cell. A dense filamentary system is visible in the apical marginal sections of the cell. The filamentary system extends into the zonula occludens of the cell membranes and into the interior of the microvilli, on the cell surface. Transmission electronmicroscopy.
× 48.000

———————————————————————————————————→

Fig. 8. Three-dimensional reconstruction of the cytoskeleton in the bronchial epithelial cells. Bundles of microfilaments permeate the cytoplasma from the nuclear membrane up to the cell membrane, and end in the contact zones of the cell membranes. The filaments form a ring-like apical marginal system that sends filamentary bundles into the microvilli. Besides, there is a network of microtubules that runs from the nuclear membrane up to the contact zones of the cell membrane, is characterized by nodular connections and can be traced up to the tips of the microvilli. The microtubules surround the basal bodies of the cilia in a latticed fashion and are inserted in their lateral processes. The tubules run on the inner side of the cell membrane, in the region of the zonulae adhaerentes
(drawing by Gerhard Puchner)

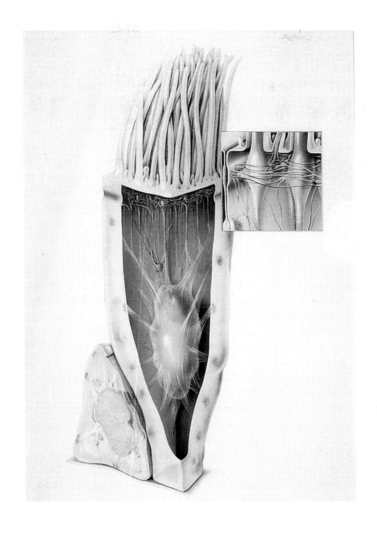

Conclusion

A morphological substrate may be attributed to the altered reactivity of the bronchial mucosa. This morphological substrate indicates grave changes in the diffusive balance of the bronchial epithelium. The altered reactivity is probably due to changes in the nerval and autonomic regulatory mechanisms of the mucosa. The efferent and afferent nerve endings in the epithelium also change. They are exposed because of the intraepithelial edema and become accessible for irritants that are inhaled, and dissolved into the sol phase of the secretory layers. The disrupted intercellular connections hinder horizontal spread of information and disturb the coordination of mucosal functions. The synchronization between the production of secretion (that takes place in a graded fashion) and the frequency of the ciliary beat could be interrupted. This causes the increased secretion to collect in the bronchial lumen. It is not transported continuously any more. Disorders in the structure of the cytoskeleton are probably responsible for the defective intake and processing of information in the single cell.

Examination of the cytoskeleton after viral infection could deliver an explanation for the link between viral infection and hyperreactivity of the mucosa. An assessment of this region reveals that coordination and synchronization of mucosal function depend very much on the structure of the cytoskeleton and this intracytoplasmic system is probably the morphological basis for the intake and intracellular transfer of information.

The close relation between the autonomic and nerval regulatory mechanisms in the bronchial mucosa offers an explanation for the rapid adjustment of mucosal function to environmental changes. The basic function is probably controlled by autonomic regulatory mechanisms and is modifed by nerval influences. The close relation between these two systems, their simultaneous changes and the subsequent morphological changes explain the hypersensitivity of the mucosa. This hypersensitivity forms a suitable background for an asthmatic attack, even from marginally irritative substances.

Clinical observations have proven that viral infection and the development of bronchial hyperreactivity are closely connected to each other. It is possible that the infection and the involvement of epithelial cells cause a more or less prolonged change in metabolism, which leads to destruction of the cytoskeleton. Proper regeneration of these filamentary and tubular cytoplasmic structures is not possible because of the grave metabolic disorders. A reversal of these changes probably depends on whether and how the virus can be rendered inactive.

Acknowledgement

We thank Sujata Wagner for translation of the manuscript from German to English.

References

1. Andres KH, v Düring (1985) Rezeptoren und nervöse Versorgung des bronchopulmonalen Systems. In: Ulmer W (Hrsg) Rezeptoren und nervöse Versorgung des bronchopulmonalen Systems. Gedon und Reuss, München
2. Morgenroth K, Donner U (1985) Elektronenmikroskopische Befunde an Brochusbiopsien zur nervösen Versorgung der menschlichen Bronchialschleimhaut. In: Ulmer W (Hrsg) Rezeptoren und nervöse Versorgung des bronchiopulmonalen Systems. Gedon und Reuss, München
3. Morgenroth K (1992) Morphologie der vagalen Versorgung. In: Kummer F (Hrsg) Das cholinerge System der Atemwege. Springer, Wien New York
4. Philippou S, Sommerfeld HJ, Wiese M, Morgenroth K (1993) The morphological substrat of autonomic regulation of the bronchial epithelium. Virchows Arch A Pathol Anat 423: 469–476

Correspondence: Prof. Dr. K. Morgenroth, Ruhr-Universität Bochum, Medizinische Fakultät, Abteilung für allgemeine und spezielle Pathologie, Universitätsstraße 150, D-44780 Bochum, Deutschland.

Discussion

Dr. Len Poulter: I have two questions. The first one is, were the samples that you've studied obtained from patients with diagnosed asthma?

Dr. Morgenroth: These were clinically stable asthmatics with proven bronchial hyperreactivity.

Dr. Poulter: Thank you. The second question is, do you have any comparative observations from other inflammatory lung disease which is not asthmatic? Demonstrating whether or not similar changes are associated with inflammation in general within the bronchial wall, or whether these changes are specific for asthma?

Dr. Morgenroth: In the course of my talk I didn't go into this detail. It was a larger group of patients, of whom we had biopsies but we focused on 12 who had asthma and bronchial hyperreactivity. We had the intention to get a differentiation between the behaviour of asthmatic bronchial mucosa and non-specific usual upper respiratory tract infection. There was no inflammatory reaction of the bronchial mucosa. These were not infected, stable asthmatics with no signs of infection in normal microscopy, there were no inflammatory signs. So they were totally intact on primary examination. But it is quite obvious, what obstacles there are to get a study like this on the way, because it's difficult to identify a virus as RSV virus or as adenovirus, by specific antibodies to begin with. And then to get the right patients and to produce these model reactions that take place under natural conditions. And further experiments are being planned in consequence of these initial experiments to initiate inflammation by cytokines and other parts of the network of inflammation in the explanted bronchial mucosa.

Bronchial hyperreactivity in the presence and absence of inflammation (A clue to asthma pathogenesis)

L. W. Poulter[1] and *C. M. Burke*[2]

[1] Department of Clinical Immunology, Royal Free Hospital, School of Medicine, London and the Asthma Research Centre
[2] James Connelly Memorial Hospital, Dublin

Summary

Asthma is now seen as a chronic inflammatory disease characterised by accumulations of T cells, antigen presenting cells, and eosinophils in the bronchial wall. To the patient however the problem is one of pathophysiology. Although spirometric readings can document airway obstruction in asthmatics these measurements may be variable and with treatment fall within the normal range during stable periods of disease. Bronchial hyperreactivity as measured by the effect on FEV_1 of inhalation of graded doses of histamine ($PC20FEV_1$) is seen to reflect disease severity. It is manifest in all asthmatics and is demonstrable even during stable periods of disease. Correlations have been observed between the extent of immunopathology in asthma and this pathophysiological parameter. However this correlation is not absolute as many examples of bronchial inflammation without bronchial hyperreactivity can be found. The current consensus that in asthma inflammation leads to bronchial hyperreactivity may therefore not be correct. Observations reported here that up to 33% of the normal population exhibit bronchial hyperreactivity but have no bronchial inflammation would support this. Indeed evidence can be found of heightened suppression of immunological reactivity in BHR + normal subjects. Far therefore, from being caused by inflammation in normal individuals the presence of bronchial hyperreactivity is associated with a down regulation of the local immune system.

BHR and bronchial inflammation may therefore be seen as parallel processes which may indeed exacerbate each other if they are coincidental but no initial cause and effect relationship may exist. Thus the development of asthma may be dependent quite simply on this coincidence. This being the case, the crucial issue for the immunopathologists rests with the reasons for the overstimulation of the T cell response.

Zusammenfassung

Bronchiale Hyperreaktivität mit und ohne Inflammation (Ein Schlüssel zur Pathogenese des Asthmas). Asthma wird heute als eine chronisch entzündliche Erkrankung angesehen, welche durch die Anhäufung von T-Zellen, Antigen präsentierenden Zellen und Eosinophilen in der Bronchialwand charakterisiert ist. Für den Patienten selbst und seine Behandlung liegt der Schwerpunkt auf dem Verständnis der Pathophysiologie. Zwar können wir mittels Spirometrie die Obstruktion als solche bei Asthmatikern dokumentieren, doch werden diese Messungen variabel sein, ja in stabilen Perioden der Behandlung sogar in den Normbereich fallen. Die bronchiale Hyperreaktivität, welche durch graduelle Steigerung von inhalativen Histamindosen und deren Auswirkung auf den FEV_1 ($PC20FEV_1$) gemessen wird, wird als Maßstab für die Schwere der Erkrankung herangezogen. Sie ist bei allen Asthmatikern vorhanden und kann sogar in stabilen Perioden der Erkrankung nachgewiesen werden.

Es wurden Korrelationen zwischen dem Ausmaß der Immunpathologie des Asthmas und diesem pathophysiologischen Parameter beobachtet.

Allerdings ist diese Beziehung keine absolute, zumal eine Reihe von Patienten eine beträchtliche bronchiale Entzündung ohne bronchiale Hyperreaktivität zeigen.

Zwar ist man übereingekommen, daß die Asthmaentzündung zur bronchialen Hyperreaktivität führen muß, doch dürfte dies nicht ganz korrekt sein. Wir haben beobachtet, daß bis zu 33% der Normalpopulation eine bronchiale Hyperreaktivität ohne gleichzeitige bronchiale Entzündung aufweisen. In der Tat gibt es Befunde, die für eine gesteigerte Suppression der immunologischen Reaktivität bei Normalpersonen mit bronchialer Hyperreaktivität sprechen. Daher ist das Vorhandensein einer bronchialen Hyperreaktivität wohl mit einer Abschwächung des lokalen Immunsystems bei Normalpersonen verquickt, jedoch weit entfernt von einer ursächlichen Entzündung.

Die bronchiale Hyperreaktivität und die bronchiale Entzündung dürften daher als parallele Prozesse anzusehen sein, welche tatsächlich zur gegenseitigen Verstärkung neigen, wenn sie gemeinsam auftreten, jedoch ohne ein initiales Wirkprinzip darzustellen. Darum ist die Entwicklung des Asthmas wohl sehr einfach auf die Koinzidenz beider Faktoren zurückzuführen. Wenn dies der Fall ist, dann bleibt als kritischer Faktor für den Immunpathologen die Beschäftigung mit der Überstimulation der T-Zell-Antwort.

Introduction

Over the last 2–3 years it has become apparent that the underlying immunopathology of bronchial asthma is a chronic T-cell dominated inflammation of the bronchial wall [1, 2].

Although the classic clinical symptoms of the disease – intermittent reversible bronchospasm – may well be associated with acute immunologic events, it is now felt that such aberrations are only manifested against the background of chronic inflammation in the mucosa of the major airways.

This inflammation is composed predominantly of a mononuclear cell infiltrate within the bronchial wall with accumulations of activated T lymphocytes, the majority of which express the phenotype CD45Ro; collections of macrophages, including a high proportion of antigen presenting cells [3] and increased numbers of eosinophils [4].

Other histopathological features within the bronchial wall of asthmatics associated with this inflammation include some variable disruption of the bronchial epithelium and a variable degree of thickening beneath the basement membrane of this epithelium [5]. Although the aetiological events promoting this inflammation remain unknown, it is felt that an antigenic stimulous promotes activation of T lymphocytes expressing predominantly a TH_2 like cytokine repertoire and the release of soluble factors from these cells causes the recruitment and activation of eosinophils which may then release soluble substances potentially damaging to the linings of the airways [6, 7].

It has been observed that such inflammation is a common feature of all asthmatics, irrespective of age, atopic status or clinical presentation. Because of the diversity of presentation within the asthmatic population resulting from these differences, the relationships between this inflammation and the physiological aspects of asthma have been difficult to establish. Although there is now no doubt that asthma is a chronic inflammatory disease (see above) the problems perceived by the patient and addressed by the clinician are physiological! Thus to understand the pathogenesis of asthma the links between immune dysfunction, inflammation, and physiological abnormality must be established.

Physiology

Investigation of physiology in the lung function laboratory reveals that asthmatics may exhibit reduced FEV_1, which may usually be reversed with bronchodilators (DELTA FEV_1); reduced FEF25–75, and other features associated with airflow obstruction [8]. These parameters however are highly variable and with many patients there may be long stable periods of disease where these parameters fall within the normal range. Measurement of peak expiratory flow rate PEFR may also reveal abnormalities particularly in its variability throughout the day, (a greater than 20% variability may be considered diagnostic for asthma) with marked reductions recorded in the mornings [9]. It is also accepted however that with treatment spirometric values may be sustained within the normal range for prolonged periods. It is thus difficult using such analyses to gauge the severity of the disease. In some cases life threatening symptomatic episodes can occur in patients showing relatively normal values.

It is also important to note that none of these tests record or predict the susceptibility of the patient to any acute stimulus which may provoke bronchospasm. On the other hand a test that is now widely accepted as monitoring this susceptibility is the test for bronchial hyperreactivity, (BHR).

Originally designed and described by Cockroft in 1977 [10], the basis of this investigation is to cause the patient to inhale histamine in graded increasing concentrations and between each "challenge" to measure the effect of such challenge on airway obstruction. This is achieved by recording the drop in FEV_1, after each progressive dose administered. In this way the provocative concentration or dose of histamine that needs to be inhaled to cause a 20% fall in FEV_1, can be accurately calculated for each patient. This is thus documented as PC20 FEV_1. It was found that in asthmatic subjects a 20% fall in FEV_1 could be achieved with a concentration of 8 mg or less of histamine while normal subjects required greater concentrations. Since its initial description this method has been used by many groups to record the bronchial hyperreactivity of asthmatic subjects. With some modification including the use of methacholine [11] as a stimulant and more recently the use of inhalation of cold air to promote broncho constriction, it is now seen as a relatively reliable and consistent method of recording airway responsiveness. It has three major advantages over other physiological tests. Firstly, it is readily controlled by the investigator and quantifiable. Secondly, it is reproducible showing limited variability over time (Table 1, Ref. 12) and thirdly, and, perhaps most importantly, BHR as detected in this way has been shown to be present in all asthmatics, again irrespective of age, atopic status or clinical presentation [11, 13]. It is also significant to note that treatment causing clinical improvement in asthmatics is reflected in improvement in this parameter [14]. Because of the consistency of this hyperreactivity in all asthmatics, and the quantifiable nature of this test, it clearly offers a

Table 1. Reproducibility of PC$_{20}$ FEV_1 (mg/ml histamine) in clinically healthy subjects

Case no	Baseline	Two weeks	Four weeks
1	3.8	4.7	3.2
2	20	12.6	> 16[a]
3	3.2	2.8	6.3
4	22.9	25.1	12.6
5	8.7	20	25.1
6	26.3	20.9	25.1

[a] Regression curve too flat to derive accurate figure

pathophysiological parameter against which the relevance of the underlying bronchial inflammation in asthma can be measured. It could be argued that the pathogenesis in this disease may indeed rest with the relationship between the immunopathology as described in the introduction and the pathophysiology which may be quantified by measurement of bronchial hyperreactivity (PC20 FEV_1).

Relationships between bronchial hyperreactivity and inflammation in asthmatics

Studies in our laboratory demonstrating a correlation between $PC20FEV_1$ and levels of HLADR expression by the inflammatory cells in the bronchmucosal of asthmatics revealed for the first time that immunopathology in terms of T cell mediated inflammation and pathophysiology may be related [2]. Since then, several studies have revealed links between these two parameters. Increasing numbers of eosinophils, T cells, and activated T cells all measured in the bronchial wall of asthmatics have been shown to correlate with BHR in these subjects [15]. Such observations have prompted investigations of the relationship between inflammation and other measures of pathophysiology in asthmatics. For example, it has recently been shown in this laboratory that in a group on non-atopic asthmatics strong correlations can be found between the levels of T cell infiltration and the FEV_1 expressed by these subjects [16].

The demonstration of relationships between bronchial hyperreactivity and mucosal inflammation have led to the conclusion that the levels of hyperresponsiveness expressed by these patients may be a direct consequence of the levels of inflammation present within the bronchial wall. The fact that *both* these parameters are consistent features of all asthmatics even when in a stable form of the disease adds weight to this argument. That is, BHR appears more related to inflammation than symptomatology. Although this evidence supports the idea that BHR is a consequence of the chronic inflammation, alternative arguments can be put forward.

Although relationships between this immunopathological and pathophysiological events appear strong within asthmatics, outside of this patient group can be found examples where extensive chronic inflammation of the bronchial wall exists but no BHR can be demonstrated. For example, only a proportion of patients with bronchiectasis, (a clinical condition where extensive inflammation of the bronchi exists) exhibit bronchial hyperreactivity [17]. This is also true in patients with bronchial eosinophilia [18] and in patients exhibiting bronchial inflammation as a result of lung transplantation [19]. In these situations patients have extensive inflammation often associated with increased levels of T lymphocyte activation but no BHR appears to emerge. Such observations clearly shed doubt on a simple cause and effect relationship between chronic inflammation in the bronchial mucosa and the emergence of this pathophysiological entity.

Bronchial hyperreactivity in normal subjects

Although the text books describe BHR as being a cardinal feature of asthma, [8] it is recognised that this physiological abnormality may not be specific for this disease (see above). BHR as defined as an exaggerated response to histamine inhalation is known to be found in a proportion of the healthy non-asthmatic population [10]. The currently accepted cut off point between the normal population and those defined as BHR+ve is 8 mg/ml histamine concentration. Those people responding to 8 mg/ml histamine or less with a 20% fall in their FEV_1 would be defined as BHR+ve whereas those failing to respond in this way to this concentration would be considered to be part of the normal population. Although this cut off point is widely used by many laboratories a more recent publication from the originators of this method has indicated that a concentration of 8 mg/ml histamine may be too high [20]. With this possibility in mind we have recently undertaken a study of 27 clinically healthy subjects within our laboratory who presented with no history of any clinically significant lung disease and were in fact being treated for sports injuries within the orthopaedic clinic. Performing $PC20FEV_1$ tests on these individuals we found that 9 of these 27 subjects exhibited BHR in that their $PC20FEV_1$ was ≤ 8mg/ml. This observation showing that up to 33% of the normal population may respond at this dose suggests two things –

1. The response point defining BHR+ve status may need to be redefined

2. BHR in any given subject might only be identified when a *change* from base line values within this individual is seen.

To determine whether or not such hyperreactivity was associated with any level of bronchial inflammation all 27 subjects underwent fibreoptic bronchoscopy during which bronchial biopsies were removed. These biopsies were frozen and cryostat sections investigated with monoclonal antibodies to reveal the distribution of T cells, antigen presenting cells, and the level of expression of HLADR. The results revealed that none of these subjects exhibited any evidence of inflammation within the bronchial wall and all figures for the numbers and distribution of these cellular elements and the levels of expression of

Table 2. Immunohistological analysis in clinically healthy subjects with and without bronchial hyperresponsiveness. Values are medians (ranges)

Group	T cells (no/10^4 μm^2)	RFD1 + cells (no/10^4 μm^2)	HLA-DR (relative optical density)
Bronchial hyper-responsiveness:			
Present (n = 9)	0.2 (0–3.3)	0.1 (0–0.69)	0.04 (0–0.04)
Absent (n = 18)	0.56 (0–1.2)	0.34 (0–2.8)	0.06 (0–0.17)
p value	NS	< 0.05	NS

Group	RFD1+	RFD7+	RFD1+ RFD7+
Total (n = 14)	11 (3–19)	44 (17–65)	46 (27–75)
Bronchial hyper-responsiveness:			
Present (n = 5)	7 (3–13)	30 (17–42)	64 (48–75)
Absent (n = 9)	11 (4–19)	54 (41–65)	34 (27–44)
p value[a]	NS	< 0.05	< 0.001

In each case median (range) of all positive cells with specific phenotype is given as a percentage:

$$\frac{\text{Specific phenotype}}{((RFD1+) + (RFD7+) + (RFD1 + RFD7+))} \times 100$$

[a] Comparing those with bronchial hyperresponsiveness with those without

class II MHC antigen within the bronchial wall fell within normal limits (Table 2). When the BHR+ group were compared to the BHR- subjects it was revealed that in those BHR+ve subjects fewer T cells, antigen presenting cells, and lower expression of HLADR were seen in the mucosa when compared to the 18 BHR-ve subjects (Table 2). Further immunofluorescence studies investigating the relative proportions of antigen presenting cells, mature phagocytes and suppressive macro-phages revealed significant differences between BHR+ve normal sub-jects and those not exhibiting bronchial BHR. When these two groups were compared a significantly greater proportion of macrophages sup-pressive macrophages was recorded in the BHR+ve group and concur-rent with this a decrease in the proportion of cells with phenotype or phagocytes was recorded (Table 2). In summary, BHR+ve normal subjects had lower numbers of T cells and antigen presenting cells, yet increased proportions of suppressive macrophages than those subjects not exhibiting bronchial hyperreactivity. *Thus BHR can exist in the absence of inflammation and is associated with reduced immunologic activity.*

The integrated picture

The results above demonstrate beyond any reasonable doubt that the presence of bronchial hyperreactivity within an individual is not de-pendent on that individual expressing a chronic inflammation within the bronchial wall. This is not to say that were bronchial hyperreactivity to be present this would not be exacerbated by the development of inflam-mation. It would indicate however that both bronchial hyperreactivity and chronic inflammation within the bronchial wall may develop inde-pendent of each other and not represent sequential events. These sugges-tions are in keeping with those of Chapman et al. [21] who have previously suggested that bronchial hyperreactivity and inflammation are parallel but not necessarily sequential events. The results from the normal subjects also demonstrate that the presence of bronchial hyper-responsiveness does not necessarily lead to symptoms of asthma!

As there are many examples, (some identified above), where chronic inflammation may be present in the bronchial wall yet again no symp-toms associated with asthma are manifest, it would also seem likely that chronic inflammation alone cannot promote the clinical situation defined as asthma. The point is not however that chronic T cell mediated inflammation and BHR are both present in all asthmatics. If, as is indicated by the results described in above, BHR is not sequential to the development of chronic inflammation, one is left to conclude that asthma will only manifest itself when both chronic inflammation and bronchial hyperreactivity happen to occur simultaneously.

Asthma may thus be defined as a failure to regulate T cell mediated immuno reactivity within the bronchial wall of the hyperreactive individual. This suggestion leaves open the question of where does the failure in T cell regulation occur?

There is to date no firm answer to this question. However in our normal individuals the BHR+ subjects showed *increased* proportions of *suppressive* macrophages and did not have asthma; whereas in the asthmatic subjects the inflammatory reactions are characterised by reduced proportions of suppressive macrophages! Macrophage regulation of T cell reactivity in the lung has been described previously [22, 23]. Whether aberrations of such mechanisms form the basis of T cell reactivity in asthma exist remain unclear. One is left to postulate how-

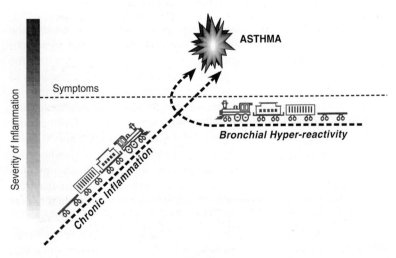

Fig. 1

ever that the coincidence of chronic inflammation and bronchial hyper-reactivity within the same individual is like two trains steaming towards each other. The inevitable crash can only be avoided by sufficient regulation of the T cell response possibly by suppressive macrophages which could turn off the inflammation (Fig. 1).

References

1. Laitinen LA, Heino M, Laitinen A, Kava T, Haahtela T (1985) Damage of the airway epithelium and bronchial reactivity in patients with asthma. Am Rev Res Dis 131: 599–606
2. Poulter LW, Power C, Burke C (1990) The relationship between bronchial immunopathology and hyperresponsiveness in asthma. Eur Respir J 3: 792
3. Hutter C, Poulter LW (1992) The balance of macrophage subsets may be customised at mucosal surfaces. FEMS Microbiol Immunol 105: 309–316
4. Djukanovic R, Wilson JW, Britten LM, Wilson SJ, Walls AF, Roche WR, Horwath PH, Holgate ST (1990) Quantitation of mast cells and eosinophils in the bronchial mucosa of symptomatic atopic asthmatics and healthy control subjects using immunohistochemistry. Am Rev Respir Dis 142: 863–871
5. Jeffrey PK, Godfrey RW, Adelroth E, Nelson F, Rogers A, Johansson SA (1992) Effects of treatment on airway inflammation and thickening of basement membrane reticular collagen in asthma. A quantitative light and electron microscopic study. Am Rev Respir Dis 145: 890–899
6. Wierenga EA, Snoek M, DeGroot C, Chretien I, Bod JD, Jansen HM, Kapsenberg M (1990) Evidence for compartmentalization of functional subsets of CD4+ T lymphocytes in atopic patients. J Immunol 144: 1651–1656
7. Montefort S, Herbert CA, Robinson C, Holgate ST (1992) The bronchial epithelium as a target for inflammatory attack in asthma. Clin Exp Allergy 2: 511–520
8. Woolcock AJ (1988) Asthma. In: Murray JF, Nadal JA (eds) Textbook of respiratory medicine. WB Saunders, Philadelphia, pp 1030–1068
9. Turner-Warwick M (1977) On observing patterns of airflow obstruction in chronic asthma. Br J Dis Chest 71: 73–86
10. Cockroft DW, Killian DN, Mellon JJA, Hargreave FE (1977) Bronchial reactivity to inhaled histamine: a method and clinical survey. Clin Allergy 7: 235–243
11. Hargreave FE, Ryan G, Thomson NC, O'Byrne PN, Latimer K, Juniper Ef et al (1981) Bronchial responsiveness to histamine or methacholine in asthma: measurement and clinical significance. J Allergy Clin Imunol 68: 347–355
12. Power C, Sreenan S, Hurson B, Burke C, Poulter LW (1993) Distribution of immunocompetent cells in the bronchial wall of clinically healthy subjects showing bronchial hyperresponsiveness. Thorax 48: 1125–1129

13. Poulter LW, Janossy G, Power C, Sreenan S, Burke C (1994) Physiological/ immunological relationships in asthma: potential regulation by lung macrophages. Immunology Today (in Press)
14. Woolcock AJ, Yan K, Salome CM (1988) Effect of therapy on bronchial hyperresponsiveness in the long-term management of asthma. Clin Allergy 18: 165–176
15. Laitinen LA, Laitinen A (1988) Mucosal inflammation and bronchial hyperreactivity. Eur Respir J 1: 488–489
16. Debenham P, Sreenan S, Burke CM, Poulter LW (1994) Relationships between lung physiology and bronchial immunopathology in asthma (submitted)
17. Pang J, Chan HS, Sung JY (1989) Prevalance of asthma, atopy and bronchial hyperreactivity in bronchiectasis: a controlled study. Thorax 44: 948–951
18. Spry CJF (1988) in Eosinophils. Oxford, University Press, p 272
19. Clelland CA, Higgenbottam TW, Stewart S, Scott JP, Wallwwark J (1990) The histological changes in transbronchial biopsy after treatment of acute lung rejection in heart lung transplantation. J Pathol 161: 105–121
20. Cockroft DW (1990) Bronchial hyperresponsiveness: In: Mygind N, Pipkorn U, Dahl R (eds) Rhinitis and asthma. Munksgaard, Copenhagen, pp 172–187
21. Chapman ID, Foster A, Morley J (1993) The relationship between inflammation and hyperreactivity of the airways in asthma. Clin Exp Allergy 23: 168–171
22. Holt PG, McMenamin C, Schon-Hegrad MA, Strichland D, Nelson D, Wilkes L, Bilyk N, Oliver J, Holt BJ, McMenamin PG (1991) Immunoregulation of asthma: control of T lymphocyte activation in the respiratory tract. Eur Respir J 4 [Suppl 13]: 6–15
23. Spiteri MA, Poulter LW (1991) Characterisation of immune inducer and suppressor macrophages from the normal human lung. Clin Exp Immunol 83: 157

Correspondence: Dr. L. W. Poulter, Department of Clinical Immunology, Royal Free Hospital, School of Medicine, Pond Street, London NW 3 2 QG.

Discussion

Dr. Lane: Do you think in terms of anti-inflammatory treatment, it's sufficient to convert asthmatics just into dirty normals?

Dr. Poulter: In terms of anti-inflammatory treatment, then I think you have little alternative but turn them into dirty normals. I think you would agree with me that there is very limited evidence, how corticosteroids for example, can bring bronchial hyperreactivity back to what one might consider a normal situation. So I think you are left with the situation, you suggest, yes.

Dr. Lane: In the normal subjects who exhibited bronchial hyperresponsiveness, in which you showed an increased number of suppressor macrophages, have you any functional correlates in terms of antigen presentation and T-cell responsiveness?

Dr. Poulter: We are pursuing in vitro investigations of this population. These macrophages appear to produce the highest levels of TGF beta. If this is the case, one could postulate that a lot of this would prevent the promotion of the differentiation of T-cells to the TH2 like phenotype, thus reducing IL4, as we are well aware that TGF beta regulates IL4 production. So if you do not have, a sufficient complement of this macrophage population, there is evidence (not yet accepted for publication, but there is evidence) that the consequence of that could be over-production of IL4.

Dr. Lane: And secondly, with reference to the study designe, have you measured bronchial hyperresponsiveness preoperatively in any of the patients, or were all measurements performed postoperatively?

Dr. Poulter: Right. The bronchial hyperreactivity was done one month after the operation. I have no evidence as to whether or not the operative procedure in the 27 that we used for comparitive study, were influenced in terms of their PC 20. I just don't know. But I would be very surprised if 4 weeks after surgery it had a significant effect on their bronchial reactivity.

N.N.: Len, have you studied any asthmatics who have been successfully treated to improve their clinical symptoms? And studied the biopsies from these patients to see whether their suppressor cell is upregulated?

Dr. Poulter: Yes we have, and it is. And in particular using inhaled corticosteroids and also using theophylline. They both upregulate the proportions of the suppressive macrophages. This is also true in Crohn's colitis. There are a large number of these suppressive cells within the gut mucosa, which disappear in a patient with Crohn's. And if treated efficaciously with steroids, they return in the gut mucosa as well. So there is no doubt at all in my mind that efficacious treatment is associated with a return of this suppressive population. Although I would hesitate as yet to confirm that they are the fundamental problem within the disease.

N.N.: If you speak about various cell types, obviously from BAL, that have been isolated and screened for the expression of various binding sites for example, what about the adhesion molecules? Did you take them into account?

Dr. Poulter: We have not studied adhesion molecules ourselves. But other people certainly have got some data that suggests that adhesion molecules both on the epithelial cells and indeed on the vasculature of the bronchial mucosa are abnormal in asthmatics. And again, these have been shown in studies, demonstrating that both with corticosteroids and again with aminophylline, there was a reversion to normal of adhesion molecules. (Advances in Immunology, 1992, S 1, 323). So I think there is no doubt that some of the changes in the bronchial mucosa are caused by changing traffic of cells from the vasculature into the tissues.

N.N.: Do you have by chance an answer why should asthmatics have such a high amount of phagocytes in their lung?

Dr. Poulter: Well, they're not so high. If you look at the overall number of macrophages within the bronchial mucosa of an asthmatic, with the exception of extremely severe cases, where there is gross inflammation, the actual number of cells is not significantly greater than in the normal. What changes in terms of the macrophage populations is the distribution of functions within that overall population, so that the increase you saw in the phagocytic cells was a result of the loss of the suppressive cells, in the same situation. Now the interesting question that arises is, are we seeing a change in recruitment of cells, or is there some influence, a cytokine or some other soluble factor within the local environment which may be changing the actual phenotype and function of a phagocyte to a suppressive cell or vice versa. Can these cells actually switch function even after the've differentiated to maturity? And I have a hunch that we're going to find out they can. In fact the loss of suppressive macrophages is the reason why you see an increase in these phagocytes, because they've actually been pushed in that direction by some imbalance in the soluble factors there.

N.N.: Did you look at the eosinophils? It might be that it is the eosinophils which are causing bronchial hyperreactivity. As far as I have seen, you have looked at lymphocytes and macrophages?

Dr. Poulter: We do look at the eosinophils, I'm not very happy with our data on the eosinophils and I would defer to other people who have made it a lifetime's work of looking at the eosinophils. In the normals there are very few eosinophils that we can see. We are using frozen sections, and using frozen sections is not the ideal way to look with regular staining at eosinophils. But if you use EG1 for example, in an immunohistochemical method, the number of eosinophils is relatively low and we saw no significant difference between the bronchially hyperreactive or the bronchially non-hyperreactive normals. But I don't put up the data because I am less happy with the way in which we quantify eosinophils than other groups that have got much better methods than us.

N.N.: But then, that could perhaps explain your data, couldn't it?

Dr. Poulter: Certainly we cannot counter the possibility that the bronchial hyperreactivity in these normals is associated with an increased level of eosinophils, although we have no evidence for it. I think the very important thing however is not really what cells are in their bronchial mucosa or not but the fact that they live with their bronchial hyperreactivity perfectly happily without ever developing any signs or symptoms of asthma. I mean, to them, that's the important thing.

Mononuclear cells in the pathogenesis of bronchial asthma

St. J. Lane and *T. H. Lee*

Department of Allergy and Allied Respiratory Disorders, London

Summary

In conclusion, there is increasing evidence implicating the central role of cells of monocyte/macrophage lineage in the pathogenesis of bronchial asthma. This evidence comes from studies on peripheral blood monocytes, BAL fluid and cells and more recently, airway immuno-histochemistry. Elucidation of the mechanisms of macrophage inter-actions may eventually lead to novel approaches in anti-asthma therapy.

Zusammenfassung

Mononukleäre Zellen bei der Pathogenese des Asthma bronchiale. Unsere Untersuchungen zielen auf die Erforschung der Effekte der von Monozyten und Makrophagen gebildeten Zellprodukte ab, welche auch die Funktion der Granulozyten und T-Zellen beeinflussen, auf-grund der erwiesenen Aktivierung von Monozyten und Makrophagen beim Asthma.

Monozyten des peripheren Blutes von Asthma-Patienten weisen eine gesteigerte Expression von $FC_\epsilon R2$ und Komplementrezeptoren auf, welche mittels der Rosettentechnik auf Monozyten von Asthma-Patien-ten nach Allergenprovokation erfaßt werden kann.

Die Alveolarmakrophagen von Patienten mit Asthma zeigen eine gesteigerte Fähigkeit zur Produktion von Eicosanoidmediatoren und

Superoxydanion. Weiters tragen diese Zellen FC$_\varepsilon$R2 auf ihrer Oberfläche und können damit zur Freisetzung von Mediatoren durch IgE-abhängige Ereignisse angeregt werden.

Die Analyse der bronchoalveolaren Lavageflüssigkeit von Patienten mit Asthma nach Antigenprovokation zeigte, daß die sekretorischen Prozesse der Makrophagen durch die Allergenprovokation aktiviert wurden.

Der Überstand einer Zellkultur aus peripheren mononukleären Zellen von atopischen Asthmatikern erhielt mehr überlebenssteigernde Aktivität für Eosinophile als der von nicht atopischen Individuen. Dieser Faktor wurde als GM-CSF ausgewiesen. Die Gegenwart von GM-CSF in der Lunge dürfte eine wichtige Rolle bei der Amplifikation der eosinophilen Entzündung spielen.

In den Bronchialbiopsien von 16 asthmatischen Patienten und 6 Normalpersonen zeigte sich, daß bei den Asthmatikern die Gesamtzahl der infiltrierenden Makrophagen gesteigert war, wobei viele dieser Zellen die phänotypischen Charakteristika der Monozyten des peripheren Blutes aufwiesen. HLA II Antigen wurde von den infiltrierenden Zellen und den Atemwegsepithelien exprimiert. Es fand sich bei den Asthmatikern eine signifikante Steigerung der aktivierten Eosinophilen, aber nicht der Neutrophilen, ferner eine signifikante Steigerung der Zahl der T-Lymphozyten mit sehr wenigen B-Lymphozyten. Diese Resultate lassen vermuten, daß die Lungenmakrophagen eine Rolle bei der Entstehung der chronischen immunmediierten Entzündung spielen, und daß eine Heterogenität von Makrophagensubpopulationen vorliegt. Wir haben zeigen können, daß die Zahl der GM-CSF bildenden Zellen in der Schleimhaut von Asthmatikern 7 × größer ist als bei Normalpersonen, wofür hauptsächlich die deutlich vermehrten Makrophagen verantwortlich sind. Zusätzlich zur GM-CSF-Exprimierung der Entzündungszellen der Schleimhaut findet sich eine sehr starke Expression dieses Zytokins auch aus den Epithelzellen.

Wenn Beclomethason in einer Dosis von 1000 µg täglich durch 8 Wochen verabreicht wurde, resultierte dies in einer signifikanten Verminderung der GM-CSF Expression im Epithel, welche überdies mit einer Steigerung des FEV$_1$ bzw. einer Zunahme der Carbachol-Provokationsschwelle korrelierte.

Eosinophile, die mit dem Überstand von Alveolarmakrophagenkulturen der Asthmatiker inkubiert und dann mit A23187 stimuliert wurden, zeigten eine Steigerung ihrer Fähigkeit, LTC$_4$ zu sezernieren (mittlere Steigerung 169 ± 37%, n = 31).

Die Hauptkomponente, die hier aktiv war, war offensichtlich eng verwandt mit dem GM-CSF.

GM-CSF und Interferon-Gamma (IFN-Gamma) steigern die Expression von HLA Klasse II Molekülen und die Antigenpräsentation. Autologe T-Zellen wurden in vitro bei normalen und asthmatischen Individuen untersucht. Die Monozyten von normalen und atopischen Personen präsentierten Recall-Antigene gegenüber T-Zellen besser als gegenüber Makrophagen.

Die gesteigerte Fähigkeit von Atemwegsmakrophagen, die T-Zellen von Asthmatikern zu aktivieren, spricht für das Vorhandensein einer unreifen Makrophagenpopulation in den Atemwegen. Es fand sich auch eine Korrelation zwischen der Lymphozytose in der BAL und der relativen Fähigkeit der Alveolarmakrophagen von Asthmatikern, Recall-Antigene zu präsentieren.

Kortikosteroide sind eine wirksame Therapie für das Bronchialasthma. Sie versagen aber bei einer kleinen Gruppe von Patienten, die zu schwerem Asthma neigen, in der Regel für längere Zeitabschnitte arbeitsunfähig sind und in der Langzeittherapie beträchtliche Schwierigkeiten bereiten. Die Steroidresistenz ist verquickt mit einer reduzierten Bindungsaffinität des Glukokortikoidrezeptors (GR) an DNA und eine verminderte Zahl von translozierten Kern-GRs, die für die DNA-Bindung verfügbar sind. Zusätzlich ist sie in vitro und in vivo mit einem Defekt in der Funktion der mononukleären Zellen verquickt.

Die Funktion der mononukleären Zellen beim Asthma kann an Monozyten des peripheren Blutes, an Alveolarmakrophagen, an der bronchialen Histologie und bei Untersuchung von Patienten mit Kortikosteroidresistenz studiert werden.

Introduction

There is accumulating evidence implicating cells of the mononuclear phagocyte lineage in the pathogenesis of bronchial asthma from functional studies on peripheral blood monocytes, analysis of the fluid and cellular phases of bronchoalveolar lavage (BAL) samples, and more recently, from histology obtained from endobronchial biopsy in vivo.

The pathology of bronchial asthma demonstrates a multicellular process. The airway mucosa is infiltrated with both mononuclear cells and granulocytes, of which the eosinophil is particularly prominent. In order to attempt an elucidation of the cellular biology of airways inflammation, it is important to understand both the interactions between different cells and the biology of each individual cell type.

Peripheral blood monocytes

In 1975 Capron discovered evidence for IgE receptors on macrophages by implicating the role of IgE antibodies in macrophage-dependent cytotoxic damage to parasites [1]. This receptor is similar, if not identical, to the $FC_\varepsilon R$ found on T and B lymphocytes and is referred to as $FC_\varepsilon R_2$ as it differs in both structure and function from the $FC_\varepsilon R_1$ found on mast cells and basophils [2]. The most important functional difference is that the $FC_\varepsilon R_2$ on macrophages is of a lower affinity (i.e. an estimated K_d for monomeric IgE binding to macrophages and U937 cells of 10 μM) and is preferentially activated by immune complexes [3, 4]. Approximately 5–10% of monocytes and alveolar macrophages (AM) bear $FC_\varepsilon R_2$ [2, 5–7]. This number rises to 20% in mildly atopic asthmatic subjects and can be downregulated by corticosteroid treatment [8]. Monocytes from asthmatic subjects have enhanced complement receptor expression (CRE) in the presence or in the absence of casein, as compared to normal controls [9]. Furthermore, the numbers of monocytes that form rosettes with complement-coated sheep erythrocytes are increased in asthmatic subjects after allergen bronchoprovocation, but not after histamine-induced bronchoconstriction [10]. Activated monocytes generate a number of proinflammatory molecules which can influence the activity of other cells types [11, 12]. For instance, supernatants from activated monocytes modulate arachidonic acid metabolism by the cyclooxygenase and lipoxygenase pathway in macrophages, fibroblasts and granulocytes [3, 13–18]. Furthermore, they activate granulocytes for certain cytotoxic functions [19–23]. For instance LPS-stimulated monocyte supernatants were capable of augmenting the antibodydependent cytotoxic killing of Schistosoma mansoni larvae by human eosinophils [24–26]. At low antibody concentrations, control medium-treated eosinophils only adhered in small numbers to the larvae and killed 2–10% of the schistosomula. In contrast, eosinophils treated with conditioned media adhered in large numbers to schistosomula and demonstrated high helmintho-toxicity (40–80%). The supernatants themselves were not toxic for the larvae and were shown to mediate these effects via enhanced eosinophil degranulation.

Monocyte-derived supernatants from atopic subjects have been shown to increase eosinophil viability to approximately 67% at 7 days as opposed to only 15% in the absence of supernatant [27]. The supernatants of atopic asthmatic individuals contain more of the viability-enhancing activity than non-atopic individuals (Fig. 1). On separation of the supernatant on C18 Sep-Pak columns, a major eosinophil viability-enhancing activity was eluted in the aqueous fraction, suggesting that it was as hydrophilic molecule. Neutralisation by specific antibodies indicated that the activity was completely inhibited by antiserum to granu-

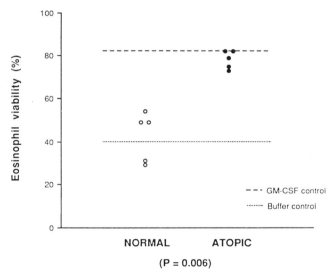

Fig. 1. Comparison of PBMC-derived culture supernatants from 5 atopic individuals with an eosinophilia of 14% ± 2% (mean ± SEM) with those from from 5 nonatopic individuals with no eosinophilia for their capacity to enhance eosinophil viability with the same target eosinophils in the assay. The negative and positive controls are incubated with medium alone and with 1ng/ml of rhGM-CSF, respectively

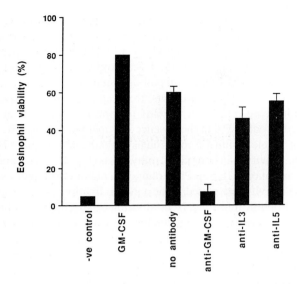

Fig. 2. The effects of different antibodies on the activity of the major eosinophil viability-enhancing activity separated by C-18 Sep-Pak fractionation. The negative control is incubation with medium alone, and the positive control is incubation with 1 ng/ml of GM-CSF

locyte-macrophage colony-stimulating factor (GM-CSF) and was unaffected by antibodies specific to IL-3 or IL-5 (Fig. 2). A second, minor viability-enhancing activity can be found in the 100% methanol fraction indicating the presence of a more hydrophobic molecule.

GM-CSF is an acidic glycoprotein with a pI of 4.5 and a molecular weight of 22.000 Daltons (D). It is eluted from size exclusion columns with an apparent molecular weight of between 15.000 and 40.000 D due to variations in its glycosylation and from anion exchange between 0.10 and 0.20 M NaCl [28]. The glycoprotein stimulates the proliferation and differentiation of normal granulocytic and monocytic stem cells and modulates the function of mature granulocytes leading to enhancement of expression of granulocyte functional antigens 1 and 2, Mo 1, Leu M5 and C3bi and enhances the function of antigen presenting cells [29–32].

GM-CSF induces histamine release from basophils and enhances eosinophil survival in culture [33, 34]. Thus, the presence of GM-CSF in the lung may precondition eosinophils for enhanced pro-inflammatory functions upon subsequent stimulation and either alone, or in concert with other cytokines, may lead to eosinophil colony formation from bone marrow progenitors. Furthermore, GM-CSF may play an important role in the amplification of the eosinophilic inflammation, which is characteristic of asthmatic airways.

Alveolar macrophages (AM)

Alveolar macrophages (AM) constitute the majority of cells recovered by BAL both in normal and asthmatic subjects [35]. Studies on BAL specimens have shown that cultured AM have an increased respiratory burst as detected by chemiluminescence when compared to normal controls [36]. Analysis of BAL fluid from asthmatic patients following antigen challenge reveals increased amounts of β-glucuronidase, whereas corresponding macrophage intracellular levels were decreased, as compared to normal controls [37, 38]. This suggests that the macrophage secretory process may be activated by allergen. Metzger has demonstrated that the total number of peroxidase positive AM obtained from BAL fluid are increased at 48 and 96 hours after allergen challenge suggesting that a population of monocytes had entered the lung from the local vascular compartment [39]. Rankin and others have demonstrated that normal AM could be activated by monoclonal IgE and specific antigen to generate both LTC_4, LTB_4 and platelet-derived growth factor [40–44].

AM supernatants derived from asthmatic patients have been shown to enhance the capacity of eosinophils to secrete LTC_4 [45]. AM supernatants derived from normal individuals had no enhancing effects when compared with culture medium. Enhancement was maximal when eosinophils were preincubated with a 1 : 6 dilution of AM supernatants for 5 minutes at 37°C and then stimulated with 5 μM A23187 for 5 minutes at 37°C. There was an inverse correlation between the % enhancement and the baseline LTC_4 production. Partial purification of the enhancing activity by HPLC on a TSK G3000 SW column revealed that it had a molecular size of approximately 30 kD and the activity was eluted in two consistent peaks at 0.17 M and 0.2 M NaCl from anion exchange HPLC (TSK DEAE 5 PW column, pH 7.4). The activities were distinct from IL-1β and TNFα. The major activity which eluted at 0.2 M NaCl was further resolved by reverse phase-high pressure liquid chromatography (RP-HPLC) on a C18 Spherisorb column with a slope gradient of 0 to 100% acetonitrile. A single peak of activity was eluted at 41% acetonitrile. The activity was inhibited by trypsin digestion and heat, and was neutralised by incubation with specific antibodies to human GM-CSF (Fig. 3). This suggested that the major active component is identical or closely related to GM-CSF. This finding is supported by the observation that pretreatment of eosinophils with recombinant GM-CSF primed the cells for enhanced LTC_4 generation following stimulation with A23187 or unopsonised zymosan. The priming may be mediated by augmented release of arachidonic acid from membrane phospholipids caused by enhancement of the rate of membrane depolarisation induced by subsequent stimulation [46–48]. The enhancement was

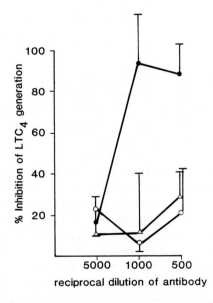

Fig. 3. The antibodies used were directed against GM-CSF (closed circles), G-CSF (open circles) or, as a control, serum obtained from animals prior to immunisation with GM-CSF was used (open triangle). Each point represents mean ± SEM (n = 4) except for anti-GM-CSF at 1 : 5000 (n = 2) and anti G-CSF at 1 : 5000 (n = 2)

dose-dependent and the maximal enhancement of LTC_4 occurred at 1 ng/ml GM-CSF for A23187 stimulation and at 10 ng/ml GM-CSF for stimulation with unopsonised zymosan. The presence of GM-CSF in the lung may lead to induction and maintenance of the eosinophil inflammatory state, characteristic of asthmatic airways, through enhanced eosinophil survival and eosinophil colony formation from bone marrow progenitors and may condition eosinophils for enhanced inflammatory functions.

The existence of low affinity $FC_{\varepsilon}R_2$, suggests that macrophages and their monocyte precursors may be directly involved in allergic responses in addition to their role in presenting antigens to T lymphocytes [5, 49]. Monocytes and macrophages are necessary for the generation of cell mediated immune responses by processing and presenting antigen to antigen specific lymphocytes, committing them to antigen specific division and cytokine generation [50]. Human AM from normal subjects function poorly for the presentation of recall antigens to autologous peripheral blood T lymphocytes and this may be important for the appropriate control of immune and inflammatory processes within the lung [51–55].

GM-CSF and IFNγ increase HLA class II molecule expression and enhance antigen presentation. Since GM-CSF production is increased in asthma the capacities of peripheral blood monocytes and macrophages obtained by bronchoalveolar lavage to present antigens to autologous T cells in vitro have been assessed in normal and atopic asthmatic individuals [56]. Monocytes presented the recall antigens, tuberculin protein purified derivative (PPD) and streptokinase-streptodornase (SKSD), to T cells better than macrophages in most of the individuals studied, as assessed by [^{3}H]-thymidine uptake. This may be due to the recognised inhibitory activity of macrophages on T cell proliferation [57]. In the asthmatic group, a number of individuals demonstrated greatly augmented macrophage antigen presenting capability relative to that of peripheral blood monocytes [56]. Aubas has previously suggested that airway macrophages of asthmatic patients have a less inhibitory activity on T cell proliferation than normal subjects [58]. Since monocytes present antigen better than macrophages and have less inhibitory activity on T cell proliferation, the increased efficacy of airway macrophages to activate T cells in asthmatic patients is consistent with the presence of an immature macrophage population in the airways. The possibility that the enhanced antigen presenting capability of the macrophage population seen in certain patients with bronchial asthma may be due to the presence of increased numbers or function of dendritic cells has not yet been evaluated.

There was a correlation between bronchoalveolar lavage lymphocytosis and the relative ability of the alveolar macrophages to present recall antigens as compared to monocytes in asthmatic, but not in normal subjects [56] (Fig. 4). The finding that airway lymphocytosis correlated with accessory cell function can be explained in at least two ways. In the first place, the presence of an alveolar lavage macrophage population, with abnormally enhanced accessory cell function, may be responsible for the infiltration of increased numbers of activated T cells in asthmatic airways. Much evidence suggests that lymphocyte localisation to the lung may be controlled by signals such as locally produced cytokines in the lung, synthesized by both the lung parenchyma and pre-existing leucocytes [59]. Alternatively, the elaboration of a hitherto uncharacterised factor from lung lymphocytes may alter the function of resident macrophages. Evidence exists for the recruitment of distinct lymphocyte subsets to the lungs, with over-representation of certain T cell receptor arrangements [60]. Furthermore, recently-activated lymphocytes also preferentially migrate into the bronchoalveolar lavage compartment of the lung and have been found in bronchial biopsies of asthmatic patients [61, 62].

There was no such correlation between the lavage lymphocytosis and accessory cell function for the presentation of PPD and SKSD in 14 patients with sarcoidosis, a condition which is also characterised by enhanced alveolar macrophage accessory cell function and pulmonary lymphocyte infiltrates [63]. Thus, while it is possible that a local subset of activated T cells may be responsible for enhancing alveolar macrophage function, this is seen only in asthma and not in sarcoidosis. Whether the relationship between the alveolar macrophage accessory cell function in lymphocytes in asthmatic individuals can be explained by infiltration into the asthmatic lung of functionally distinct subsets of site- and allergenspecific CD4 positive cells, with a specific pattern of cytokine release, remains to be determined [64].

Workers have isolated phenotypically and functionally distinct macrophage subpopulations from human and rat BAL fluid [56, 65, 66]. Using the monoclonal antibodies RFD1 and RFD7 they can discrimi-

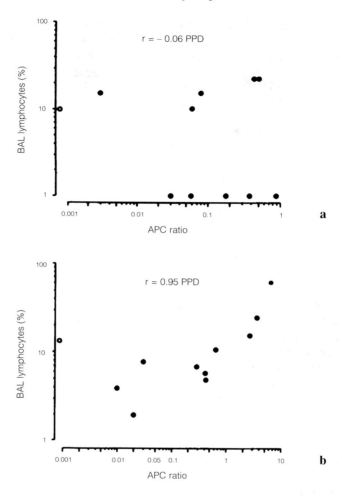

Fig. 4. Correlation between percentage lymphocytosis in bronchoalveolar lavage (BAL) and the antigen presenting cell (APC) ratio in normal (*a*) and in asthmatic (*b*) subjects for purified protein derivative (PPD) and streptokinase-streptodornase (SKSD). The APC ratio is the ratio of macrophage : monocyte antigen presenting capability. Each point is an individual subject, and the open circles indicate an APC ratio of zero. The correlation coefficient (r) was calculated by linear regression analysis

nate between phagocyte (RFD7+), antigen presenting cells (RFD1+), and suppressor macrophages (RFD1+RFD7+) [67]. All three subpopulations have been identified within the alveolar macrophage pool of humans [68]. The significance of the presence of the three subsets is highlighted by the observation that their proportions within BAL altered dramatically with the presence of lung inflammation and that T cell responsiveness might be influenced by relative changes within this heterogeneous AM population [69].

Bronchial histology

Post mortem studies of the airways of patients who died from severe asthma showed epithelial damage, plugging of the airway lumen by cellular debris, mucosal oedema, thickening of the epithelial basement membrane, smooth muscle hypertrophy and intense inflammatory cell infiltrate in the bronchial wall. More recent data obtained from biopsy specimens of human bronchus confirm the presence of airway inflammation in subjects with mild to moderately severe asthma, suggesting that the inflammation of the bronchi is an important factor in the mechanism of asthma [70–78]. Biopsies from asthmatic subjects demonstrated an increased activated heterogeneous submucosal macrophage infiltration when compared with that of non-asthmatic controls indicating that lung macrophages may have a role to play in the mechanism of the chronic immune-mediated response seen in the airway mucosa of the asthmatic subjects [79] (Fig. 5).

The presence of high numbers of inflammatory cells, such as eosinophils and macrophages, in the bronchial submucosa of asthmatic individuals, might be explained by increased levels of cytokines such as GM-CSF. Using polyclonal and monoclonal antibodies against GM-CSF for immunohistochemistry of airway biopsies obtained by fibreoptic bronchoscopy, it has been shown that there are 7-fold greater numbers of GM-CSF staining cells in the bronchial submucosa of asthmatic patients than in that of normal subjects [80] (Fig. 6). Approximately two thirds of cells staining for GM-CSF are macrophages and approximately three quarters of macrophages stain positively for GM-CSF. In addition to the staining of submucosal inflammatory cells for GM-CSF, the epithelial cell stains very strongly for this cytokine.

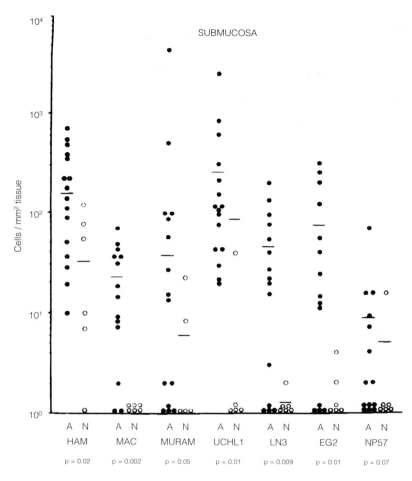

Fig. 5. Individual cell counts in the submucosa for each of the antibodies used on biopsies obtained from asthmatic (A) and from normal subjects (N). HAM = HAM56: pan-macrophage marker; MAC387 = monoclonal anti-monocyte monclonal antibody; MURAM = polyclonal anti-muramidase(lysozyme); UCHL1 = anti-memory T-cell monoclonal antibody; LN3 = anti-HLA-DR monoclonal antibody; EG2 = anti-eosinophil cationic protein monoclonal antibody; NP57 = anti-neutrophil elastase monoclonal antibody. The p values given were obtained using the Mann-Whitney U test

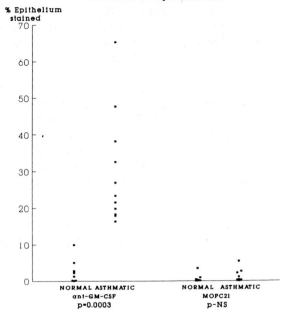

Fig. 6. Percentage of airway epithelium stained with a monoclonal anti-GM-CSF or control mouse IgG₁ myeloma protein MOPC21 in 11 asthmatic and 9 normal subjects

The extent of staining, as quantitated by the hue saturation-intensity (HSI) colour image analysis, indicated that asthmatic airway epithelial cells stained more strongly with a polyclonal or a monoclonal antibody to GM-CSF than that of normal subjects. Administration of inhaled beclomethasone dipropionate 1000 µg daily for 8 weeks resulted in a significant reduction in GM-CSF expression in the airway epithelium in patients given corticosteroids as compared to those given placebo (Fig. 7). There was a correlation between the suppression of GM-CSF staining by inhaled corticosteroids and the increase in FEV_1 and decrease in carbachol responsiveness ($r = 0.61$, $p = 0.05$ and $r = 0.8$, $p < 0.01$ respectively). These data suggest a strong involvement of GM-CSF in the pathogenesis of bronchial asthma. This might be due to GM-CSF involvement in the survival, differentiation and activation, and chemotaxis of inflammatory cells such as macrophages and eosinophils [81–85]. In addition, these findings are consistent with in vitro data demonstrating the production of GM-CSF, IL-1β, IL-6 and IL-8 by cultured airway epithelial cells in the presence and absence of stimulation by IL-1β or TNFα [86–90].

The substantial macrophage infiltration in the airways of asthmatic subjects led to a study of possible chemoattractant factors produced by the airway epithelium with a specificity towards monocytes, such as monocyte chemoattractant protein-1 (MCP-1) [91]. MCP-1 is an 8–15 kD basic peptide which is encoded by the human homologue of the mouse JE gene, the first PDGF early response gene to be identified [88, 92, 93]. It belongs to the interleukin-8 gene superfamily of small

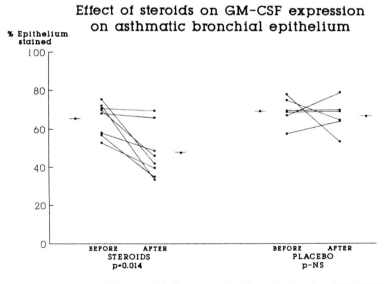

Fig. 7. Percentage of airway epithelium stained with a polyclonal anti-GM-CSF before and after inhalation of 1,000 µg beclomethasone dipropionate or placebo for 8 weeks in 8 and 6 asthmatic subjects, respectively. Mean values are indicated beside each group. Statistical differences were analysed by the Wilcoxon signed rank test for matched pairs

molecular weight cytokines known as chemokines. It belongs to the same subfamily as RANTES and MIP-1α and β, the c-c superfamily, which is distinct from the IL-8/NAP-1/C-X-C subfamily. MCP-1 is known to activate different inflammatory cells, such as monocytes and can be produced by different cell types, including monocytes, macrophages and pulmonary alveolar macrophages [94–101].

Sousa and colleagues used a monoclonal and polyclonal antibody to detect MCP-1 expression on bronchial biopsies from asthmatic and normal subjects [91]. Both antibodies demonstrated an upregulation of MCP-1 expression in the asthmatic bronchial epithelium. There was also upregulation of MCP-1 expression in the submucosa of the asthmatic bronchial biopsies, many of the positive cells having the histological characteristics of macrophages. The increased levels of MCP-1 in asthmatic airways suggest that it may play a role in macrophage recruitment and activation, and thereby contribute to the inflammatory pathology of bronchial asthma. Therefore bronchial epithelial cell produces two cytokines, namely GM-CSF and MCP-1, which demonstrate specificities towards the monocyte. The observation that IL-1β, a major cytokine produced by monocytes, can lead to upregulation of cytokine production by the epithelial cell, suggests that there may be an important amplification network between the infiltrating monocyte/macrophage and the airway epithelial cell in asthma.

References

1. Capron A, Dessaint JP, Capron M, Bazin H (1975) Specific IgE antibodies in immune adherence of normal macrophages to Schistosoma mansoni schistosomules. Nature 253: 474–475
2. Spielberg HL (1984) Structure and function of Fc receptors for IgE on lymphocytes, monocytes and macrophages. Adv Immunol 35: 61–88
3. Joseph M, Tonnel AB, Torpier G, Capron A, Arnoux B, Benveniste J (1983) Involvement of immunoglobulin E in the secretory process of alveolar macrophages from asthmatic patients. J Clin Invest 71: 221–230
4. Bach MK, Brashler JR, Hammarstrom S, Samuelsson B (1980) Identification of leukotriene C1 as a major component of slow reacting substance from rat mononuclear cells. J Immunol 125: 115–117
5. Melewicz FM, Kline LE, Cohen AB, Spiegelberg HL (1982) Characterisation of IgE receptors for IgE on human alveolar macrophages. Clin Exp Immunol 49: 364–370
6. Anderson CL, Spiegelberg HL (1981) Macrophage receptors for IgE: bind-

ing of IgE to specific IgE receptors on a human macrophage cell line U937. J Immunol 126: 2470–2473

7. Dessaint JP, Capron A, Joseph M,Bazin H (1979) Cytophilic binding of IgE to the macrophage. II. Immunologic release of lysozomal enzyme from macrophages by IgE and anti-IgE in the rat. Cell Immunol 46: 24–34

8. Melewicz FM, Zeiger RS, Mellon MH, O'Connor RD (1981). Increased peripheral blood monocytes with Fc receptors for IgE in patients with severe allergic disorders. J Immunol 126: 1592–1595

9. Kay AB, Diaz P, Carmichael J, Grant IWB (1981) Corticosteroid-resistant chronic asthma and monocyte complement receptors. Clin Exp Immunol 44: 576–580

10. Carroll MP, Durham SR, Walsh G, Kay AB (1985) Activation of neutrophils and monocytes after allergen-and histamine-induced bronchoconstriction. J Allergy Clin Immunol 75: 290–296

11. Elias JA, Schreiber AD, Gustilo K, Chien P, Rossman MD, Lammie PJ, Danielle RP (1985) Differential interleukin-1 elaboration by unfractionated and density fractionated human alveolar macrophages and blood monocytes: relationship to cell maturity. J Immunol 135: 3198–3204

12. Liu MC, Proud D, Lichtenstein LM, McGlashan DW, Schleimer RP, Adkinson LF, Kagey-Sobotka A, Schulman ES, Plaut M (1986) Human lung macrophage-derived histamine-releasing activity is due to an IgE-binding factor(s). J Immunol 136: 2588–2595

13. Dessein AJ, Lenzi HL, David JR (1983) Modulation of the cytotoxicity of human blood eosinophils by factors secreted by monocytes and T-lymphocytes. Monogr Allergy 18: 45–51

14. Dessein AJ, Lee TH, Elsas P, Ravelese J, Silberstein D, David JR, Austen KF, Lewis R (1986) Enhancement by monokines of leukotriene generation by human eosinophils and neutrophils stimulated with calcium ionophore A23187. Immunol 136: 3829–3838

15. Di Persio JF, Billing P, Williams R, Gasson JC (1988) Human granulocte-macrophage colony-stimulating factor and other cytokines prime human neutrophils for enchanced arachidonic acid release and leukotriene B_4 synthesis. J Immunol 140: 4315–4322

16. Kurland JI, Bockman RS, Boxmeyer HE, Moore MA (1978) Limitation of excessive myelopoiesis by the intrinsic modulation of macrophage-derived prostaglandin E. Science 199: 552–555

17. Nathan CF (1987) Secretory products of macrophages. J Clin Invest 79: 319–326

18. Hancock WW, Pleau ME, Kobzik L (1988) Recombinant granulocyte-macrophage colony stimulating factor down-regulates expression of IL-2 receptor on human mononuclear phagocytes by induction of prostaglandin E. J Immunol 140: 3021–3025

19. Dessein AJ, Lenzi HL, Bina JC, Carvalho EM, Weiser WY, Andrude ZA, David JR (1984) Modulation of eosinophil cytoxicity by blood mononuclear cells from healthy subjects and patients with chronic Schistosomiasis mansoni. Cell Immunol 85: 100–113

20. Vadas MA, Nicola N, Lopez AF, Metcalf D, Johnson G, Pereira A (1984) Mononuclear cell-mediated enhancement of granulocyte function in man. J Immunol 133: 202–207
21. Veith MC, Butterworth AE (1984) Enhancement of human eosinophil mediated killing of Schistosoma mansoni larvae by mononuclear cell product in vitro. J Exp Med 57: 1828–1843
22. Berger M, Wetzler EM, Wallis SR (1988) Tumour necrosis factor is the major monocyte product that increases complement receptor expression on mature human neutrophils. Blood 71: 151–158
23. Beutler B, Krochin N, Milsark IW, Luedke C, Cerami A (1986) Control of cachectin (tumor necrosis factor) sythesis: mechanisms of endotoxin resistance. Science 232: 977–980
24. Elsas P, Lee TH, Lenzi HL, Dessein LJ (1987) Monocytes activate eosinophils for enhanced helminthotoxicity and increased generation of leukotriene C₄. Ann Inst Pasteur/Immunol 138: 97–116
25. Thorne KJ, Richardson BA, Taverne J, Williamson DJ, Vadas MA, Butterworth AE (1986) A comparison of eosinophil activating factor with other monokines and lymphokines. Eur J Immunol 46: 1143–1148
26. Vadas MA, David JR, Butterworth AE, Pisani NT, Siongok TA (1979) A new method for the purification of human eosinophils and a comparison of the ability of these cells to damage schistosomula of schistosoma mansoni. J Immunol 122: 1228–1236
27. Burke LA, Hallsworth MP, Litchfield TM, Davidson R, Lee TH (1991) Identification of the major activity derived fron cultured human peripheral blood mononuclear cells, which enhances eosinophil viability, as granulocyte-macrophage colony-stimulating factor (GM-CSF). J Allergy Clin Immunol 2: 226–235
28. Nicola NA, Metcalf M, Johnson GR, Burgess AW (1979) Separation of functionally distinct human granulocyte-macrophage colony-stimulating factor. Blood 54: 614–627
29. Kapp A, Zeck-Kapp G, Donner M, Luger T (1988) Human granulocyte-macrophage colony stimulating factor: an effective direct activator of human polymorphonuclear neutrophilic granulocytes. J Invest Dermatol 91: 49–55
30. Morrisey PJ, Bressler L, Park LS, Alpert A, Gillis S (1987) Granulocyte macrophage colony stimulating factor augments the primary antibody response by enhancing the function of antigen presenting cells. J Immunol 139: 1113–1119
31. Sullivan R, Griffin JD, Simons ER, Schafer AI, Meshulaas T, Fredette JP, Maas AK, Gadenne AS, Leavitt JV, Melnick DA (1987) Effects of recombinant human granulocyte and macrophage colony-stimulating factor on signal transduction pathways in human granulocytes. J Immunol 139: 3422–3430
32. Lopez AF, Williamson DJ, Gamble JR, Begley CG, Harlan JM, Klebonoff SJ, Waltersdorf A, Wong G, Clark SC, Vadas MA (1986) Recombinant human granulocyte-macrophage colony-stimulating factor stimulates in vitro mature human neutrophil and eosinophil function, surface receptor expression and survival. J Clin Invest 78: 1220–1228

33. Haak-Frendscho M, Arai N, Arai K-I, Baeza ML, Finn A, Kaplan AP (1988) Human recombinant granulocyte-macrophage colony stimulating factor and interleukin 3 cause basophil histamine release. J Clin Invest 82: 17–19

34. Owen WF, Rothenberg ME, Silberstein DR, Gasson JC, Stevens RL, Austen KF, Soberman RJ (1987) Regulation of human eosinophil viability, density and function by granulocyte/macrophage colony-stimulating factor in the presence of 3T3 fibroblasts. J Exp Med 166: 129–141

35. Eschenbacher WL, Gravelyn TR (1987) A technique for isolated airway segment lavage. Chest 92: 105–109

36. Cluzel M, Damon M, Chanez P (1987) Enhanced alveolar cell luminol-dependent chemiluminescence in asthma. J Allergy Clin Immunol 80: 195–201

37. Tonnel AB, Joseph M, Gosset PH, Gosset P, Fournier E, Capron A (1983) Stimulation of alveolar macrophages in asthmatic patients after local provocation test. Lancet 1 (8339): 1406–1408

38. Murray JJ, Tonnel AB, Brash AR, Roberts LJ, Gosset P, Workman R, Capron A, Oates JA (1986) Release of Prostaglandin D2 into human airways during acute allergen challenge. N Engl J Med 315: 800–804

39. Metzger WJ, Zavala D, Richerson HB, Moseley P, Iwamoto P, Monick M, Sjoerdsma K (1987) Local allergen challenge and bronchoalveolar lavage of allergic asthmatic lungs. Am Rev Respir Dis 135: 433–440

40. Rankin JA, Hitchcock M, Merrill WW, Bach MK, Brashler JR, Askenase PW (1982) IgE-dependent release of leukotriene C$_4$ from alveolar macrophages. Nature 297: 329–331

41. Rankin JA (1986) IgE immune complexes induce LTB$_4$ release from rat alveolar macrophages. Ann Inst Pasteur Immunol 137: 364–367

42. Martin TR, Raugi G, Merritt T, Henderson WR (1987) Relative contribution of leukotriene B$_4$ to the neutrophil chemotactic activity produced by the resident human alveolar macrophage. J Clin Invest 8: 1114–1124

43. Fels AOS, Pawlowski NA, Cramer EB, King TK, Cohn ZA, Scott WA (1982a) Human alveolar macrophages produce leukotriene B$_4$. Proc Natl Acad Sci USA 79: 7866–7870

44. Martinet Y, Rom WN, Grotendorst GR, Martin GR, Crystal RG (1987) Exaggerated spontaneous release of platelet-derived growth factory by alveolar macrophages from patients with idiopathic pulmonary fibrosis. N Engl J Med 317: 202–209

45. Howell CJ, Pujol JL, Crea AEG, Davidson R, Gearing AJH, Godard P, Lee TH (1989) Indentification of an alveolar macrophage-derived activity in bronchial asthma which enhanced leukotriene C4 generation by human eosinophils stimulated by ionophore (A23187) as granulocyte-macrophage colony-stimulating factor (GM-CSF). Am Rev Respir Dis 140: 1340–1347

46. Dahinden CA, Zingg J, Maly FE, de Weck AJ (1988) Leukotriene production in human neutrophils primed by recombinant human granulocyte/macrophage colony stimulating factor and stimulated with complement component C5a and FMLP as second signals. J Exp Med 167: 1281–1295

47. Pennington JE, Rossing TH, Boerth LW, Lee TH (1985) Isolation and partial characterization of a human alveolar macrophage-derived neutrophil activating factor. J Clin Invest 75: 1230–1237
48. Arnoux B, Duval D, Benveniste J (1980) Release of platelet activating factor (PAF-acether) from alveolar macrophages by the calcium ionophore A23187 and phagocytosis. Eur J Clin Invest 10: 437–441
49. Rosenstreich DL, Mizel SB (1978) The participations of macrophages and macrophage cell lines in the activation of T lymphocytes by mitogens. Immunol Rev 40: 102–135
50. Unanue ER, Allen PM (1987) The basis for the immunoregulatory role of macrophages and other accessory cells. Science 236: 551–557
51. Holt PG, Schon HM, Oliver J, Holt BJ, McMenamin PG (1990) A contiguous network of dendritic antigen-presenting cells within the respiratory epithelium. Int Arch Allergy Appl Immunol 91: 155–159
52. Toews GB, Vial WC, Dunn MM, Gazetta P, Navez P, Stastny P, Lipscomb MF (1984) The accessory function of human alveolar macrophages in specific T-cell proliferation. J Immunol 132: 181–186
53. Ettensohn DB, Roberts NJ (1983) Human alveolar macrophage support of lymphocyte responses to mitogens and antigens: analysis and comparison with autologous peripheral blood-derived monocytes and macrophages. Am Rev Resp Dis 128: 516–522
54. Mackaness GB (1971) The induction and expression of cell-mediated hypersensitivity in the lung. Am Rev Respir Dis 104: 813–828
55. Rich EA, Tweardy DJ, Fujiwara H, Ellner JJ (1987) Spectrum of immunoregulatory functions and properties of human alveolar macrophages. Am Rev Respir Dis 136: 258–265
56. Gant V, Cluzel M, Shakoor Z, Rees PJ, Lee TH, Hamblin A (1992) Alveolar macrophage accessory cell function in bronchial asthma. Am Rev Respir Dis 146: 900–904
57. Yarborough WC Jr, Wilkes DS, Weissler JC (1991) Human alveolar macrophages inhibit receptor-mediated increases in intracellular calcium concentration in lymphocytes. Am J Respir Cell Mol Biol 5: 411–415
58. Aubas P, Cosso B, Godard P, Michel FB, Clot J (1984) Decreased suppressor cell activity of alveolar macrophages in bronchial asthma. Am Rev Respir Dis 130: 875–878
59. Kelley J (1990) Cytokines of the lung. Am Rev Respir Dis 141: 765–788
60. Augustin A, Kubo RT, Sim GK (1989) Resident pulmonary lymphocytes expressing the gamma-delta T cell receptor. Nature 240: 239–241
61. Beacham CH, Daniele RP (1982) Migration of recently divided B and T lymphocytes to peritoneum and lung. Cell Immunol 74: 284–293
62. Hamid Q, Azzawi M, Ying S, Moqbel R, Wardlaw A, Corrigan C, Bradley B, Durham S, Collins J, Jeffery P, Quint D, Kay AB (1991) Expression of mRNA for interleukin-5 in mucosal bronchial biopsies from asthma. J Clin Invest 87: 1541–1546
63. Gant VA, Shakoor ZS, Barbosa IL, Hamblin AS (1991) Normal and sarcoid

alveolar macrophages differ in their ability to present antigen and to cluster with autologous lymphocytes. Clin Exp Immunol 86: 494–499

64. Kapsenberg ML, Wierenga EA, Bos JD, Jansen HM (1991) Functional subsets of allergen-reactive human CD4 +ve cells. Immunology Today 12: 392–395

65. Spiteri MA, Clarke SW, Poulter LW (1988) Phenotypic and functional changes in alveolar macrophages contribute to the pathogenesis of pulmonary sarcoidosis. Clin Exp Immunol 74: 359–364

66. Spiteri MA, Poulter LW (1991) Characterization of immune inducer and suppressor macrophages from the normal human lung. Clin Exp Immunol 83: 157–162

67. Poulter LW, Campbell DA, Munro C, Janossy C (1986) Discrimination of human macrophages and dendritic cells by means of monoclonal antibodies. Scand J Immunol 24: 351–357

68. Spiteri MA, Clarke SW, Poulter LW (1992 a) Isolation of phenotypically and functionally distinct macrophage subpopulations from human bronchoalveolar lavage. Eur Resp J 5: 717–726

69. Spiteri MA, Clarke SW, Poulter LW (1992 b) Alveolar macrophages that suppress T cell responses may be crucial to the pathogenetic outcome of pulmonary sarcoidosis. Eur Respir J 5: 394–403

70. Beasley R, Roche WR, Roberts JA, Holgate ST (1989) Cellular events in the bronchi in mild asthma and after bronchial provocation. Am Rev Respir Dis 139: 806–817

71. Jeffery PK, Wardlaw AJ, Nelson FC, Collins JV, Kay AB (1989) Bronchial biopsies in asthma: an ultrastructural, quantitative study and correlation with hyperreactivity. Am Rev Respir Dis 140: 1745–1753

72. Jeffrey PK, Godfrey RW, Adelroth E, Nelson F, Rogers A, Johanson S-A (1992) Effects of treatment on airway inflammation and thickening of basement membrane reticular collagen in asthma. A quantitative light and electron microscopic study. Am Rev Respir Dis 145: 890–899

73. Bousquet J, Chanez P, Lacoste JY, Barneon G, Ghavanian N, Enander I, Venge P, Ahlstedt S, Simony-Lafontaine J, Godard P, Michel FB (1990) Eosinophilic inflammation in asthma. N Engl J Med 323: 1033–1039

74. Bousquet J Chanez P, Campbell AM, Lacoste JY, Poston R, Enander I, Godard P, Michel FB (1991) Inflammatory processes in asthma. Int Arch Allergy Immunol 94: 227–232

75. Djukanovic R, Roche WR, Wilson JW, Beasley CRW, Twentyman OP, Howarth PH, Holgate ST (1990 a) Mucosal inflammation in asthma. Am Rev Respir Dis 142: 434–457

76. Azzawi M, Bradley B, Jeffery PK, Frew AJ, Wardlaw AJ, Knowles G, Assoufi B, Collins JV, Durham S, Kay AB (1990) Identification of activated T lymphocytes and eosinophils in bronchial biopsies in stable atopic asthma. Am Rev Respir Dis 142: 1407–1413

77. Laitinen LA, Laitinen A, Haahtela T (1993) Airway mucosa inflammation even in patients with newly diagnosed asthma. Am Rev Respir Dis 147: 697–704

78. Busse WW, Calhoun WF, Sedgwick JD (1993) Mechanism of airway inflammation in asthma. Am Rev Respir Dis 147: 20–24
79. Poston RN, Chanez P, Lacoste JY, Litchfield T, Lee TH, Bousquet J (1992) Immunohistochemical characterization of the cellular infiltration in asthmatic bronchi. Am Rev Respir Dis 145: 918–921
80. Sousa AR, Poston RN, Lane SJ, Nakhosteen JA, Lee TH (1993) GM-CSF expression in bronchial epithelium of asthmatic airways: decrease by inhaled corticosteroids. Am Rev Respir Dis 147: 1557–1561
81. Metcalf D (1985) The granulocyte-macrophage colony-stimulating factors. Science 229: 16–22
82. Ruel C, Coleman DL (1990) Granulocyte-macrophage colony-stimulating factor: pleiotropic cytokine with potential clinical usefulness. Rev Infect Dis 12: 41–62
83. Tai PC, Spry CJ (1990) The effects of recombinant granulocyte-macrophage colony-stimulating factor (GM-CSF) and interleukin-3 on the secretory capacity of human blood eosinophils. Clin Exp Immunol 80: 426–434
84. Wang JM, Colella S, Allavela P, Mantovani A (1987) Chemotactic activity of human recombinant granulocyte-macrophage colony-stimulating factor. Immunology 60: 439–444
85. Grabstein KH, Urdal DL, Tushinski RJ, Mochizaki DY, Price VL, Cantrell MA, Gillis S, Conlon PJ (1986) Induction of macrophage tumoricidal activity by granulocyte-macrophage colony stimulating factor. Science 232: 506–508
86. Marini M, Soloperto M, Mezzetti M, Fasoli A, Mattoli S (1991) Interleukin-1 binds to specific receptors on human bronchial epithelial cells and up-regulates granulocyte-macrophage colony-stimulating factor synthesis and release. Am J Respir Cell Mol Biol 4: 519–524
87. Marini M, Vittori E, Hollemborg J, Mattoli S (1992) Expression of the potent inflammatory cytokines, granulocyte-macrophage colony-stimulating factor and interleukin-6 and interleukin-8, in bronchial epithelial cells of patients with asthma. J Allergy Clin Immunol 89: 1001–1009
88. Soloperto M, Mattoso VL, Fasoli A, Mattoli S (1991) A bronchial epithelial cellderived factor in asthma that promotes eosinophil activation and survival as GM-CSF. Am J Physiol 260 (Lung Cell Mol Physiol 4): L530–L538
89. Cromwell O, Hamid Q, Corrigan C, Barkans J, Meng Q, Collins P, Kay AB (1992) Expression and generation of interleukin-8, IL-6 and granulocyte-macrophage colony stimulating factor by bronchial epithelial cells and enhancement by IL1-β and TNFα. Immunology 77: 330–337
90. Mattoli S, Miante S, Calabro F, Mezzeti M, Fasoli A, Allegra L (1990) Bronchial epithelial cells exposed to isocyanates potentiate activation and proliferation of T cells. Am J Physiol (Lung Cell Mol Physiol 3) 259: 1320–1327
91. Sousa AR, Lane SJ, Nakhosteen JA, Yoshimura T, Lee TH, Poston RN (1994) Increased expression of the monocyte chemoattractant protein-1 in bronchial tissue from asthmatic subjects. Am J Respir Cell Mol Biol 10(2): 142–147

92. Leonard EJ, Yoshimura (1990) Human monocyte chemoattractant protein-1 (MCP1). Immunology Today 11: 97–101
93. Rollins BJ, Walz A, Baggioline M (1991) Recombinant human MCP-1/JE induces chemotaxis, calcium flux, and the respiratory burst in human monocytes. Blood 78: 1112–1116
94. Jiang V, Beller Dl, Frendl G, Graves DT (1992) Monocyte chemoattractant protein-1 regulates adhesion molecule expression and cytokine production in human monocytes. J Immunol 148: 2423–2428
95. Yoshimura T, Yunk N, Moore SK, Appela E, Lerman Ml, Leonard EJ (1989) Human monocyte chemoattractant factor-1 (MCP-1) Full length cDNA cloning, expression in mitogen-stimulated blood mononuclear leukocytes, and sequence similarities to mouse competence gene JE. FEBS Lett 244: 487–493
96. Colotta F, Borre A, Wang JM, Tattaneli M, Maddalena AF, Polentarutti N, Perri G, Montovani A (1992) Expression of monocyte chemotactic cytokine by human mononuclear phagocytes. J Immunol 148: 760–765
97. Martin CA, Dorf ME (1991) Differential regulation of interleukin-6, macrophage inflammatory protein-1 and JE\MCP-1 cytokine expression in macrophage cell lines. Cell Immunol 135: 245–258
98. Antoniades HN, Neville-Golden J, Galanopoulos T, Kradin RN, Valente AJ, Graves DT (1984) Expression of monocyte chemoattractant protein-1 mRNA in human idiopathic pulmonary fibrosis. Proc Nat Acad Sci USA 89: 5371–5375
99. Brieland JK, Jones ML, Clarke SJ, Baker JB, Warren JS, Fantone JS (1992) Effect of acute inflammatory lung injury on the expression of monocyte chemoattractant protein-1 (MCP-1) in rat pulmonary alveolar macrophages. Am J Respir Cell Mol Biol 7: 134–139
100. Denholm EM, Wolber FM, Phan SH (1989) Secretion of monocyte chemotactic activity by alveolar macrophages. Am J Pathol 135: 571–580
101. Cochron BH, Leonard EJ, Stiles CD (1983) Molecular cloning of gene sequences regulated by platelet-derived growth factor. Cell 33: 939–947

Correspondence: Prof. T. H. Lee, Department of Allergy and Allied Respiratory Disorders, 4th Floor, Hunt's House, Guy's Hospital, London SE1 9RT, U.K.

Discussion

Dr. Poulter: I have two questions. When you were looking at your eosinophils, you started off with an in vitro assay showing a relationship between GM-CSF and eosinophils. Then you moved to the lavage. Now I was sort of expecting to see the comparison being made with the bronchial biopsies. Have you had a chance yet to look at the relationships between the monocytes and the GM-CSF actually in the tissues? And the levels of eosinophils there, which would be a sort of a natural conclusion.

Dr. Lane: We have double-stained bronchial submucosal biopsies with the macrophage marker HAM-56 and with anti GM-CSF. We found that the macrophage population in asthmatic subjects was substantially increased in number from non-asthmatic control subjects. In addition, we found the macrophages to be activated, to comprise a heterogenous population and a greater proportion of the asthmatic individuals were positive for GM-CSF. In all the asthmatic subjects studied there was an increase in eosinophilic infiltration as compared to non-asthmatic controlled subjects. We have not correlated the degree of infiltrating macrophages, infiltrating eosinophils and clinical symptoms. In addition, the percentage of macrophages secreting GM-CSF was increased in the asthmatic population. This would certainly be an interesting correlation and indeed would be consistent with in vitro data on peripheral blood monocytes.

Dr. Poulter: Is there any relationship between the proportions of the epithelium expressing GM-CSF and the increases in monocytes that you described in the tissues?

Dr. Lane: GM-CSF is an important cytokine in asthma as its expression in the bronchial epithelium is greatly increased in asthmatic subjects and its expression is decreased in vivo by treatment with corticosteroids which correlates with clinical improvement. We have not yet correlated submucosal monocyte infiltration with epithelial GM-CSF expression although this would be an interesting relationship to establish.

Dr. Poulter: In the tissues, do you think it's GM-CSF coming from the monocytes, which is actually on the epithelium, or do you think the epithelial cells are making GM-CSF rather than the monocytes that are actually infiltrating the lamina propria?

Dr. Lane: There are many in vitro data showing that cultured epithelial cells derived from asthmatic individuals actively secrete GM-CSF into the cultured supernatants to a greater degree than non-asthmatic control subjects. In addition, GM-CSF mRNA has been shown to be expressed by bronchial epithelium both in vivo and in cultured human epithelial cells in vitro. Hence the epithelium is also capable of expressing and secreting GM-CSF. In view of this capacity of the epithelial cells, it is highly probable that the protein visualised by immunohisto-chemistry is being produced by the bronchial epithelium.

Kummer: I'm very impressed that the epithelial cell is much more potent in immune modulation and in the network of cytokines than we originally thought.

N.N.: How specific is a morphologic assessment of the eosinophil viability from your point of view?

Dr. Lane: We have assessed eosinophil viability as the percentage of cells which stain positively with Trypan blue. Trypan blue exlusion is a well described morphologic change consistent with cell death. It is important to remember that these experiments compared the effects of the presence or absence of monocyte supernatants on the eosinophil viability and are thus controlled.

N.N.: Did you look at IL8 as far as the chemotactic response of mononuclear cells is concerned?

Dr. Lane: Epithelial cells secrete IL8 but we haven't looked at its expression in the bronchial mucosa. This would be very interesting.

N.N.: That increased number of dendritic cells, as you say, may have to do with a local expansion and expression of GM-CSF or IL8 maybe.

Dr. Lane: I agree that dendritic cells may be an important source contributing to production of the cytokines such as GM-CSF and IL8. Cells of the monocyte macrophage lineage are not the only cell types which produce these cytokines and indeed GM-CSF is produced by T cells and IL8 is produced by epithelial cells.

Epithelium under stress

W. R. Pohl[1] and D. J. Romberger[2]

[1] 2. Medizinische Abteilung, Wilhelminenspital, Wien
[2] University of Nebraska, Med. Ctr., Omaha

Summary

Cultured bovine bronchial epithelial cells respond to thermal stress with induction of HSP 70 and this stress is associated with increased fibronectin release. We conclude that epithelial cells are capable of responding to stress and that altered release of mediators involved in repair may be part of the stress response. TGF-β may modulate HSP 70 expression, to which extent, however, needs further evaluation.

Epithelial cells play an important role in inflammatory processes such asthma and chronic bronchitis having the potential to communicate with a number of immune cells in order to help direct the inflammatory response.

In this context, epithelial cells influence processes in the airways including smooth muscle tone, inflammation, repair and fibrosis. Improved understanding of the role of epithelial cells in airway diseases will provide the means for more specific direction of new therapeutic strategies.

(See glossary of abbreviations at the end!)

Zusammenfassung

Epithel unter Stress. Atemwegserkrankungen wie die chronische Bronchitis und das Asthma sind durch wiederholte Episoden der Atemwegsentzündung gekennzeichnet. Der Mechanismus bei der Reparation der entzündlichen Schäden ist heute noch weitgehend unklar, spielt aber eine Schlüsselrolle darin, ob bleibende Schäden wie die subepitheliale Fibrose sich entwickeln werden oder nicht. Moderne Konzepte deuten darauf hin, daß die Migration, Proliferation und Differenzierung der Atemwegsepithelien für die Reparation einer verletzten Oberfläche von essentieller Bedeutung sind.

Fibronektin, ein extrazelluläres Matrixglykoprotein, spielt eine bereits gut definierte Rolle bei zelluären Prozessen, welche für Reparationsmechanismen Bedeutung haben.

Hitzeschockproteine (HSPs, Streßprotein) werden in Säugetierzellen gebildet, wenn dieselben auf eine Reihe von Streßsituationen reagieren, wie etwa Hyperthermie, Hypoxie, Ischaemie und oxydative Schädigung. Bei Menschen zählt die Familie der 70 kD HSP zu den prominentesten Vertretern der Hitzeschockproteine.

HSPs sind jüngst im Atemwegsepithel gefunden worden, wo sie wohl eine weitere Schutzfunktion gegenüber Streß ausüben. In weiterer Folge konnte gezeigt werden, daß TGF-beta die Produktion von HSP vermehrt.

Introduction

Airway epithelium is a vital component and essential in providing a protective barrier from the atmosphere. However, this is not the only function of epithelial cells, there is increasing line of evidence that epithelial cells participate actively in immunological processes. Numerous studies could clarify that airway epithelial cells can be considered as important constituents of the inflammatory response modulating epithelial repair, fibrosis and airway smooth muscle tone [1, 2].

Asthma and chronic bronchitis are inflammatory disorders of the airways characterized by repeated episodes of injury leading to epithelial dysfunction. Inflammation seen in these diseases is associated with significant epithelial damage [3]. Normal repair following such an injury includes the restoration of the epithelium. Aberrant repair as documented in these diseases leads to the development of anatomical changes contributing to the clinical picture of fixed airway obstruction [4].

Mechanisms controlling airway epithelial repair are not completely understood. Bronchial epithelial cells are thought to play an important role in directing inflammatory responses in the airways. By virtue of their biological activities epithelial cells are capable of participating in restoration of the epthelium by influencing mechanisms such as attachment, migration, and proliferation [5–8].

There is evidence that epthelial cells left behind after injury interact with other parenchymal cells as well as inflammatory cells.

Inflammation

Epithelial cells represent a first line of defense with the ability to release after stimulation a number of mediators which are essential for inflammatory processes [9]. These mediators include chemotactic factors for neutrophils and monocytes such as arachidonate metabolites (including LTB4), interleukin-8, RANTES and MIP-1 [10–17]. Additionally, epithelial cells exert effects on eosinophils and macrophages via releasing GM-CSF and thereby promoting survival, activation and proliferation of these cells [18–20].

Furthermore, airway epithelial cells are capable of expressing surface molecules such as ICAM-1 which is essential to interact with leucocytes and is important to clear ongoing viral infections [21–23]. Epithelial cells can also directly interact with T-lymphocytes which addtionally may be critical for the acceleration of immune processes [24]. Thus, airway epithelial cells seem to have the potential to control and even amplify inflammatory responses.

Repair meachnisms

An important aspect of the resolution of the inflammation is that bronchial epithelial cells are capable of migration which is thought to be an early feature of repair [25]. Components of the extracellular matrix seem to be important factors influencing migration of epithelial cells across a provisional matrix after injury [26]. Fibronectin is one of these constituents participating in the early events of repair mechanisms and, furthermore, has been shown to be a potent chemotactic factor for epithelial cells [27]. Increased expression of fibronectin has been demonstrated in wound matrix and there is evidence that epithelial cells are an important source of cellular derived fibronectin [28, 29]. Shoji et al. have demonstrated that bronchial epithelial cells in culture release fibronectin which has chemotactic properties for both lung fibroblasts and other bronchial epithelial cells [30]. Thus, epithelial cell fibronectin can modulate the activity of cells in and associated with the airway epithelium.

Epithelial cell derived fibronectin may represent a form of the protein produced by either differential splicing of the gene or by post translational modifications of the molecule [31]. The production of epithelial fibronectin is modulated by a variety of substances including transforming growth factor-beta (TGF-β), a cytokine implicated in both wound repair and fibrotic processes [28, 32]. TGF-β modulates epithelial cell fibronectin in several ways causing not only an increase in production mediated through changes in fibronectin mRNA, but also targeting the release in an apical or basal fashion and stimulating the expression of specific alternative splice variants [33].

Importantly, there is evidence that TGF-β is produced by epithelial cells which may act in an autocrine or paracrine fashion to stimulate fibronectin production by these cells [34]. TGF-β influences a number of processes involved in repair including differentiation and cell growth [35]. In addition, TGF-β effects other epithelial cell functions such as morphology and attachment [32]. Several investigators demonstrated that TGF-β exposure causes squamous morphology [34]. Spurzem et al. recently reported that TGF-β also influences the attachment of bronchial epithelial cells to several extracellular matrix components including fibronectin, vitronectin and type IV collagen [36].

Stress response and heat shock proteins

The interaction between epithelial cells and extracellular matrix may be regulated by mechnism which are not completely understood and still have to be evaluated. One recent observation that might be relevant to airway repair is that epithelial cells are capable of producing heat shock proteins (HSP) [37]. These proteins are induced in a variety of cells in response to several forms of stress including hyperthermia, ischemia, and oxidative injury and are thought to confer further tolerance to stress. In humans, the 70 kD HSP family is among the most prominent classes of heat shock proteins [38, 39]. HSPs may play a major role in the pathophysiology of inflammatory diseases such as asthma and chronic bronchitis.

They are involved in cellular repair mechanisms and in protection from cellular injury. Furthermore, it has been demonstrated that TGF-β modulates HSP production [40].

In a recent study we examined whether stress of heat shock may alter airway repair mechanisms by modulation of bronchial epithelial cell fibronectin release [41].

Methods

To investigate this hypothesis, we examined whether thermal stress altered bronchial epithelial fibronectin release in vitro. As a control, we assessed the expression of heat shock protein 70, an inducible form of the heat shock protein family, in cultured bovine bronchial epithelial cells. Bovine BECs were obtained from lungs using a modification of the protease digestion method of Wu and cultured in serum free medium. Nearly confluent monolayer cultures were exposed to thermal stress and fourty eight hours later cells were harvested for SDS-PAGE and subsequent Western blotting using a monoclonal antibody to HSP 70.

Results

We could demonstrate that HSP 70 was induced by thermal stress in comparison to control cultures without thermal stress. To evaluate the release of fibronectin by thermal stress, epithelial cells were treated in the same fashion and media harvested at 48 and 72 hours after thermal stress. Media were assayed for fibronectin by ELISA. The maximal response was observed at 48 hours. There was a two-fold increase in fibronectin in thermal stressed cultures as compared to controls (thermal stressed $1091.7 + 103.1$ ng/10^6 cells/hour; mean + SEM; control $575.1 + 70.9$ ng/10^6 cells/hour mean + SEM, $p < 0.05$).

In a following study we investigated whether TGF-β is capable of inducing heat shock protein production in cultured bronchial epithelial cells. Exposure of bronchial epithelial cells to increasing amounts of TGF-β resulted in an increased induction of fibronectin m-RNA as previously described by Romberger et al. on the other hand, epithelial cells treated with TGF-β in the same fashion demonstrated a decreased expression of HSP 70 m-RNA (Fig. 1).

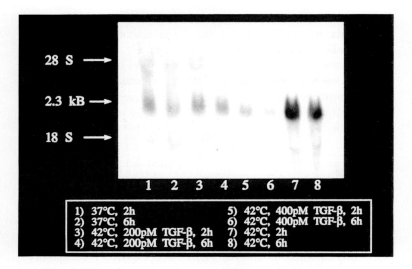

Fig. 1. Expression of heat shock protein HSP 72 in bronchial epithelial cells. HSP expression from epithelial cells is low at 37°C (1, 2). Under thermal stress, there is a marked increase of expression (7, 8), which can be attenuated by 200 pM of TBF-β (3, 4) resp. abolished by 400 pM of TGF-β (5, 6)

Discussion

These seemingly contradictory effects may be linked to different requirements for chaperoning functions or may reflect that prolonged stimulation of epithelial cells like during chronic inflammatory processes could lead to a consumption of self protective cellular mechanisms associated with decreased expression of HSP 70.

Summary

Cultured bovine bronchial epithlial cells respond to thermal stress with induction of HSP 70 and this stress is associated with increased fibronectin release. We conclude that epithelial cells are capable of responding to stress and that altered release of mediators involved in repair may be part of the stress response. TGF-β may modulate HSP 70 expression, to which extent, however, needs further evaluation.

Epithelial cells play an important role in inflammatory processes such asthma and chronic bronchitis having the potential to communicate with a number of immune cells in order to help direct the inflammatory response.

In this context, epithelial cells influence processes in the airways including smooth muscle tone, inflammation, repair and fibrosis. Improved understanding of the role of epithelial cells in airway diseases will provide the means for more specific direction of new therapeutic strategies.

Glossary of abbreviations

LTB 4 (leucotriene B$_4$)
MIP-1 (macrophage inflammatory protein-1)
GM-CSF (granulocyte macrophage colony stimulating factor)
ICAM-1 (intercellular adhesion molecule-1)
HSP (heat shock protein)
TGF-β (transforming growth factor β)

References

1. 34th Annual Thomas L (1992) Petty Aspen Lung Conference. Chest 101 (3) [Suppl]: 2–85
2. Lung Biology in Health and Disease. The Airway Epithelium. Lenfant C (ed) Marcel Dekker, New York
3. Laitinen LA, Heino M, Laitinen A, Kava T, Haahtela T (1985) Damage of the airway epithelium and bronchial reactivity in patients with asthma. Am Rev Respir Dis 131: 599–606
4. Dunnill MS (1960) The pathology of asthma, with special reference to changes in the bronchial mucosa. J Clin Path 13: 27–33
5. Lane BP, Gordon R (1974) Regeneration of rat tracheal epithelium after mechanical injury. Proc Soc Exp Biol Med 145: 1139–1144
6. McDowell EM, Ben T, Newkirk C, Chang S, DeLuca LM (1987) Differentiation of tracheal mucociliary epithelium in primary cell culture recapitulates normal fetal development and regeneration following injury in hampsters. Am J Pathol 129: 511–522
7. Rutten MJ, Ito S (1983) Morphology and electrophysiology of guinea pig gastric mucosal repair in vitro. Am J Physiol 244: 171–182
8. Moore R, Carlson S, Madara JL (1989) Rapid barrier restitution in an vitro model of intestinal epithelial injury. Lab Invest 60: 237–244
9. Jordana M, Clancy R, Dolovich J, Denburg J (1992) Effector role of the epithelial compartment in inflammation. Annals of the New York Academy of Sciences 664: 180–189

10. Churchill L, Chilton FH, Resau JH, Bascom R, Hubbard WC, Proud D (1989) Cyclooxygenase metabolism of endogenous arachidonic acid by cultured human tracheal epithelial cells. Am Rev Respir Dis 140: 449–459

11. Mattoli S, Marini M, Fasoli A (1992) Expression of the potent inflammatory cytokines, GM-CSF, IL6, and IL8, in bronchial epithelial cells of asthmatic patients. Chest 101 (3) [Suppl]: 27–29

12. Koyama S, Rennard SI, Leikauf GD, Shoji S, VonEssen S, Claassen L, Robbins RA (1991) Endotoxin stimulates bronchial epithelial cells to release chemotactic factors for neutrophils. J Immunol 147: 4293–4301

13. Holtzman MJ, Hansbrough JR, Rosen GD, Turk J. Uptake release and novel species-dependent oxygenation of arachidonic acid in human and animal airway epithelial cells. Biochem Biophys Acta 963: 401–413

14. Noah TL, Paradiso AM, Madden MC, McKinnon KP, Devlin RB (1991) The response of a human bronchial epithelial cell line to histamine: intracellular calcium changes and extracellular release of inflammatory mediators. Am J Respir Cell Mol Biol 5: 484–492

15. Heeger P, Wolf G, Meyers C, Sun MJ, O'Farrell SC, Krensky AM, Neilson EG (1992) Isolation and characterization of cDNA from renal tubular epithelium encoding murine RANTES. Kidney Intl 41: 220–225

16. Standiford TJ, Kunkel SL, Basha MA, Chensue SW, Lynch JP, III, Toews GB, Westwick J, Strieter RM (1990) Interleukin-8 gene expression by a pulmonary epithelial cell line: a model for cytokine networks in the lung. J Clin Invest 86: 1945–1953

17. Standiford TJ, Rolfe MW, Kunkel SL, Lynch III JJP, Burdick MD, Gilbert AR, Orringer MB, Whyte RI, Strieter RM (1993) Macrophage inflammatory protein-1 alpha expression in interstitial lung disease. J Immunol 151: 2852

18. Ohtoshi T, Vancheri C, Cox G, Gauldie J, Dolovich J, Denburg JA, Jordana M (1991) Monocyte-macrophage differentiation induced by human upper airway epithelial cells. Am J Respir Cell Mole Biol 4: 255–263

19. Ohtoshi T, Tsuda T, Vancheri C, Abrams JS, Gauldie J, Dolovich J, Denburg JA, Jordana M (1991) Human upper airway epithelial cell-derived granulocyte-macrophage colonystimulating factor induces histamine-containing cell differentiation of human progenitor cells. Int Arch Allergy Appl Immunol 95: 376–384

20. Marini M, Soloperto M, Mezzetti M, Fasoli A, Mattoli S (1991) Interleukin-1 binds to specific receptors on human bronchial epithelial cells and upregulates granulocyte/macrophage colony-stimulating factor synthesis and release. Am J Respir Cell Mol Biol 4: 519–524

21. Rothlein R, Czajkowski M, O'Neill MM, Marlin SD, Mainolfi E, Merluzzi VJ (1988) Induction of intercellular adhesion molecule 1 on primary and continuous cell lines by pro-inflammatory cytokines. J Immunol 141: 1665–1669

22. Wegner CD, Gundel RH, Reilly P, Haynes N, Letts LG, Rothlein R (1990) Intercellular adhesion molecule-1 (ICAM-1) in the pathogenesis of asthma. Science 2: 456–459

23. Tosi MF, Stark JM, Hamedani A, Smith CW, Gruenerts DC, Huang YT (1992) Intercellular adhesion molecule-1 (ICAM-1)dependent and ICAM-1-independent adhesive interactions between polymorphonuclear leukocytes and human airway epithelial cells infected with parainfluenza virus type 2. J Immunol 149: 3345-3349

24. Nickoloff BJ, Turka LA (1993) Keratinocytes: key immunocytes of the integument. Am J Pathol 143 (2): 325–331

25. Keenan KP, Wilson TS, McDowell EM (1983) Regeneration of hamster tracheal epithelium after mechanical injury. IV. Histochemical, immuno-cytochemical and ultrastructural studies. Virchows Arch (Cell Pathol) 43: 213–240

26. McGowan S (1992) Extracellular matrix and the regulation of lung development and repair. FASEB J 6: 2895–2904

27. Grinnell F, Billingham RE, Burgess L (1981) Distribution of fibronectin during wound healing in vivo. J Invest Dermatol 76: 181–189

28. Romberger DJ, Beckmann JD, Claassen L, Ertl RF, Rennard SI (1992) Modulation of fibronectin production of bovine bronchial epithelial cells by transforming growth factor-β. Am J Respir Cell Mol Biol 7: 149–155

29. Stoner GD, Katoh Y, Foidart JM, Trump BF, Steinert PM, Harris CC (1981) Cultured human bronchial epithelial cells: blood group antigens, keratin, collagens, and fibronectin. In Vitro 17 (7): 577–587

30. Shoji S, Ertl RF, Linder J, Romberger DJ, Rennard SI (1990) Bronchial epithelial cells produce chemotactic activity for bronchial epithelial cells. Am Rev Respir Dis 141: 218–225

31. Morle A, Zhang Z, Ruoslahti E (1994) Superfibronectin is a functionally distinct form of fibronectin. Nature 367: 193–196

32. Ruoslahti E (1988) Fibronectin and its receptors. Annu Rev Biochem 57: 375–413

33. Wang A, Cohen DS, Palmer E, Sheppard D (1991) Polarized regulation of fibronectin secretion and alternative splicing by transforming growth factor β. J Biol Chem 266: 15598–15601

34. Sacco O, Romberger D, Rizzino A, Beckmann J, Rennard SI, Spurzem J (1992) Spontaneous production of TGF–β2 by primary cultures of bronchial epithelial cells: effects on cell behavior in vitro. J Clin Invest 90: 1379–1385

35. Border W, Ruoslahti E (1992) Transforming growth factor-β in disease: the dark side of tissue repair. J Clin Invest 90: 1–7

36. Spurzem JR, Sacco O, Rickard KA, Rennard SI (1993) Transforming growth factor-beta increases adhesion but not migration of bovine bronchial epithelial cells to matrix proteins. J Lab Clin Med 122: 92–102

37. Bonay M, Soler P, Riquet M, Battesti JP, Hance AJ, Tozi A (1994) Expression of heat shock proteins in human lung and lung cancer. Am J Respir Cell Mol Biol 10: 453-461

38. Polla SP (1988) A role for heat shock proteins in inflammation. Immunology Today 9: 134–137

39. Morimoto RI (1993) Cells in stress: transcriptional activation of heat shock genes. Science 259: 1409–1410

40. Takenaka J, Hightower LE (1992) Transforming growth factor-β1 rapidly induces Hsp70 and Hsp90 molecular chaperous in cultured chicken embryo cells. J Cell Biol 152: 568–577
41. Prouse B, Pladsen P, Pohl W, Rennard SI, Romberger DJ (1994) Thermal stress induces bronchial epithelial cell heat shock protein expression and fibronectin release. Resp Crit Care Med 149: 996

Correspondence: Dr. W. R. Pohl, II. Medizinische Abteilung, Wilhelminen-spital, Montleartstraße 37, A-1171 Wien.

Discussion

Dr. Lane: Thank you very much for your very enjoyable talk. Is there any evidence that the cytokines which you demonstrated are increased in bronchial asthma have also a pathogenic role?

Dr. Pohl: Whether or not TGF beta is really of any effect in asthma, I couldn't find any paper about this. But Len just told me that they are looking for a TGF-beta modulating cytokine, and that TGF beta could play a role. What we have speculated is that TGF beta is very important for repair and that wound repair is an ongoing mechanism in asthma.

Dr. Poulter: I also enjoyed the talk very much. If you look at the epithelial cells in vitro, they are sort of isolated, whereas they're in a very specifically defined location in vivo. Do you know if there is any effect, whether the stimulus of these cells comes from the outside i.e. the airway, or from the inside? Because the surface of the cell, that sees the stimulus, would be totally different, depending on which side the stimulus may come from.

Dr. Pohl: I really don't know. But we do know that there is always some problem when you work with cultured epithelial cells. They differentiate during culture so it's not really clear in the sense that we always work with the same cell everyday. So that might be a big influence regarding these results.

Moderator: Doctor Morgenroth might have some comment on the difference between the outside layer of the epithelium and the inside layer. Doctor Morgenroth.

Dr. Morgenroth: As far as my work is concerned, I focused on the things that come from the lumen. I consider the basal membrane as non-permeable for things that could come from the inside to cause the changes which I have observed. If you do a cell culture, you should have some means of identifying what cells grow and how they differentiate, and whether or not these are the ones that you hope that really grow in the culture.

Moderator: The integralists are the pathologists. And the pathologists, as Dr. Poulter and Dr. Morgenroth look at the entire context. And they're interested in whether it's true or not, that cells that behave like this in the culture would do so in the integrity of the whole system.

St. Lane: There's one more question. Your epithelial cells are responding by producing these cytokines to various exogenous artificial stimuli. Have any experiments been done looking at coculture assays taking out macrophages from asthmatics and seeing what the supernatants will do to the epithelial cells, for example.

Dr. Pohl: I thought about this, but we haven't done it till now. But we are starting now epithelial cells gained by brush biopsies to culture through bronchoscopy. That's the only thing what we are doing on human cells at the moment. But we haven't looked at the effect of supernatants on these cells so far.

Structural and inflammatory changes in the airways of asthmatics

P. K. Jeffery

Lung Pathology, Department of Histopathology, Royal Brompton National
Heart and Lung Institute, United Kingdom

Summary

These studies support the role of airway wall inflammation in bronchial asthma, confirm the involvement of eosinophils and their degranulation and implicate, further, T lymphocytes and interleukin 5 as controllers of the eosinophilic response. Several factors may contribute to apparent airways hyperresponsiveness including fragility of surface epithelium, thickening of the airway wall, uncoupling of airway wall and parenchyma or loss of lung elastic recoil. Thickening of the airway wall may be due to bronchial vessel dilatation and oedema, or enlargement of the mass of bronchial smooth muscle or mucus-secretory glands. Each of these changes may be induced by the inflammatory cells which infiltrate the airway wall in both mild and severe asthma: their predominant cytokine products can clearly influence the characteristic profile of cell phenotypes, their activation and the chronicity of the inflammatory response. In this regard, mast cells and lymphocytes may be important initiators and controllers, whilst activated eosinophils are key reactor cells whose increased presence is associated with increased airways responsiveness.

Zusammenfassung

Strukturelle und entzündliche Veränderungen in den Atemwegen von Asthmatikern. Das tödliche Asthma zeigt histologisch drei Charak-

teristika: Die Abschilferung und Zerstörung des Epithels der Atemweg-
soberfläche, die Verdickung seiner retikulären Basalmembran und die
Entzündung der Mukosa. Wir haben die Ultrastruktur und die Immunhis-
tologie der Bronchialmukosa an Autopsiepräparaten von Personen, die
im Status asthmatikus gestorben waren (n = 15), analysiert, sowie von
Bronchialbiopsien, welche durch fiberoptische Bronchoskopie von Patienten mit relativ mildem atopischen Asthma gewonnen worden waren
(n = 33). Bei ersteren (fatale Asthmafälle) wurde ein Vergleich mit den
Befunden von plötzlich verstorbenen nicht asthmatischen Kontrollen (n
= 10) durchgeführt, bei den milden Asthmatikern wurde ein Vergleich
sowohl mit atopischen Nicht-Asthmatikern als auch normalen gesunden
Personen durchgeführt (n = 32). Beim milden Asthma korrelierte der
Verlust von Oberflächenepithel mit dem Grad der Hyperreaktivität auf
Metacholin, als Hinweis auf die Brüchigkeit des Atemwegsepithels. Die
Verdickung der retikulären Basalmembran und die Schleimhautentzün-
dung, welche beim fatalen Asthma beobachtet wurde, sind ebenso beim
milden atopischen Asthma nachzuweisen. Hier besteht ein Trend zu
größeren Zahlen von CD45+ Leukozyten, sowohl beim fatalen als auch
beim milden Asthma, wobei hier CD3+ (T) Lymphozyten und Eosinophile, aber nicht Neutrophile betroffen sind. Die Aktivitätsmarker für
Lymphozyten und Eosinophile (CD25 bzw. EG2) sind signifikant erhöht
und bei symptomatischen Patienten mit dem Auftreten von unregelmäßig
gestalteten lymphozytenähnlichen Zellen verquickt. Je größer die Zahl
der aktivierten Eosinophilen, desto größer ist auch der Grad der Hyperreaktivität. Ähnliche Veränderungen können beim intrinsischen und Be-
rufsasthma beobachtet werden. Unter Verwendung von in situ Hybridi-
sation können spezifische Zytokinsignale für IL-5 und für IL-5 und GM-
CSF mRNA nachgewiesen werden, und zwar innerhalb der Bronchialmukosa von symptomatischen atopischen Asthmatikern, bzw. in der
Folge von Allergenprovokation. Kurzfristige (4 Wochen) und langfristi-
ge Behandlung durch inhalative Kortikosteroide reduzierte die charak-
teristische Eosinophilie signifikant. Die Untersuchung der BAL bestätigt
das Vorhandensein von Eosinophilen, die Freisetzung von ECP und die
vermehrte Expression von T-Zell-Genen, mit den Codes für IL4, IL5 und
GM-CSF. Diese Ergebnisse stützen die Annahme, daß pro-inflammator-
ische Zytokine (insbesondere Interleukin 4 und 5) an der durch Eosinophile mediierten Epithelzerstörung sogar bei milden atopischen Asthmatikern beteiligt sind.

Introduction

Asthma is not considered to be a single disease entity: it is more likely a complex of conditions, all of which have in common airflow limitation. Extrinsic (allergic), intrinsic and occupational forms are often recognized but thus far cannot be distinguished by the pathologist. The limitation to airflow is usually variable over short periods of time and reversible spontaneously or with treatment, albeit an important underlying (chronic) irreversible component may persist between episodes of acute attacks [1]. Airways hyperresponsiveness (AHR) is a characteristic and common feature of asthma but is by no means unique to it. Contraction of bronchial smooth muscle is an important component contributing to reduced airflow during an attack of asthma: this is evidenced by the efficacy of inhaled β-agonists as rescue medication. But there is now increasing evidence that inflammation of the airway mucosa and thickening of the airway wall are responsible for the underlying chronicity of severe asthma and that this may contribute to AHR even in relatively mild stable disease. The following synopsis outlines the salient features of fatal and also mild asthma and highlights those features which have been implicated in the causation of AHR.

Changes in fatal asthma

Post-mortem studies of fatal asthma have shown widespread tenacious plugs in small bronchi and consequent of the lungs to retract on opening the chest [1] (Fig. 1a, b). Histochemically the airway plugs in asthma are a mixture of inflammatory exudate and mucus and included desquamated surface epithelial cells, lymphocytes and eosinophils [2, 3]. Loss of surface epithelium, thickening of the airway wall due to enlargement of bronchial smooth muscle mass, bronchial vessel dilatation, oedema (see Fig. 2), enlargement of submucosal glands and inflammation with characteristic tissue eosinophilia are usual but not universal features of fatal asthma. In subjects with a history of chronic asthma who die of non-respiratory causes, thickening of the subepithelial reticular layer (i.e. referred to as the basement membrane by light microscopy) is the most constant feature (Figs. 1b and 3) [4].

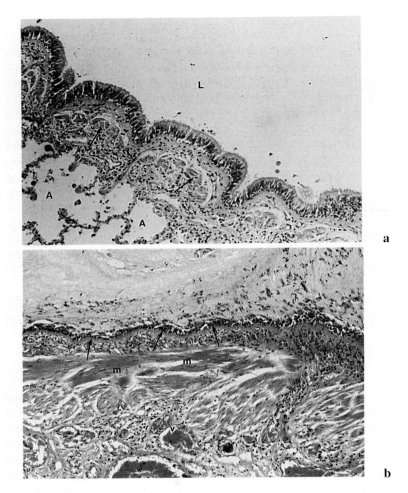

Fig. 1. H&E-stained sections of intrapulmonary bronchi: *a* from a road traffic accident death, showing intact ciliated surface epithelium with indistinct reticular basement membrane, a few immuno-competent cells and small amounts of bronchial smooth muscle (arrows). Airway lumen (*L*) and alveoli (*A*). × 120. *b* from a subject who died in *status asthmaticus* illustrating the sloughing and disruption of surface epithelium and lumenal exudate, the underlying thickened reticular basement membrane (arrows), increased numbers of inflammatory cells, enlarged mass of bronchial smooth muscle (*m*) and dilatation of bronchial vessels (v). × 120

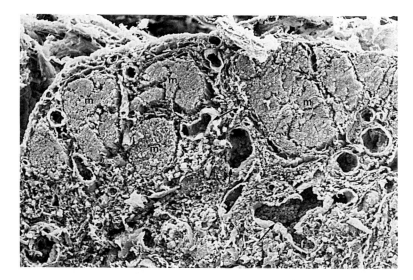

Fig. 2. Scanning electron microscopic (SEM) appearance of a fracture across the bronchial wall of a subject dying of an attack of acute severe asthma. There is extensive loss of surface epithelium, enlargement of bronchial smooth muscle (*m*) and widespread dilatation of bronchial vasculature (arrows). × 250

Surface epithelium

Histologically, shedding and damage of airway surface epithelium is prominent, both in fatal asthma and in biopsy specimens of patients with mild stable extrinsic asthma. The greater the loss of surface epithelium in biopsy specimens the greater appears to be the degree of AHR and this is suggested to reflect the extreme fragility of the epithelial lining [5, 6]. Squamous metaplasia may be seen occasionally also.

Reticular basement membrane

Thickening and hyalinization of the reticular basement membrane has long been recognized as a consistent change in asthma [see 6–9]. Whilst there may also be focal and variable thickening of the reticular basement membrane in COPD, and other inflammatory chronic diseases of the

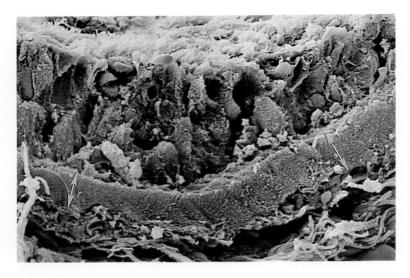

Fig. 3. SEM appearance of bronchial mucosa from a subject with 25 years history of asthma but who died of non-respiratory cause. The surface epithelium appears to be disorganized and there is homogenous thickening of its reticular basement membrane (arrows). × 2000

lung (such as bronchiectasis and tuberculosis [10], the lesion is highly characteristic and is present early on even in mild asthma [11]. The extent of thickening of the reticular basement membrane does not appear to be associated with AHR [6].

Mucus-secreting cells

Whilst not as extensive as in chronic bronchitis, there is significant submucosal gland enlargement seen in status asthmaticus [2]. Unlike bronchitis, where there is evidence of a relative loss of serous acini, the ratio of serous to mucous acini is thought to be maintained in asthma [11]. Goblet cell hyperplasia (in bronchi) and mucous metaplasia (in bronchioli) are variously reported or refuted in asthma and the role of mucus-hypersecretion in airway plugging is still unclear.

Bronchial smooth muscle

The percentage of bronchial wall occupied by bronchial smooth muscle (BSM) shows a striking and perhaps invariable increase in status asthmaticus [2, 12, 34] (see Figs. 1b and 2). In contrast, in the absence of wheeze, values for the percentage of the wall occupied by BSM in segmental bronchi in chronic bronchitis and emphysema fall largely within the normal range: intermediate values are, however, present in so-called wheezy bronchitis. The relative contributions of smooth muscle fibre hypertrophy and hyperplasia to the increase in muscle mass are unknown as is the relationship of BSM enlargement to AHR.

Inflammatory cell infiltrate

In fatal asthma there is a marked inflammatory infiltrate in both airway wall and occluding plug: eosinophils are characteristic, lymphocytes are abundant and neutrophils are notably absent [see 1–3]. The association of tissue eosinophilia and asthma is a strong one [13, 14]. Some (but not all) studies of biopsies obtained by fibreoptic bronchoscopy or at open lung biopsy in relatively mild asthma demonstrate the presence of eosinophils also [15, 16] (Fig. 4). BAL and biopsy data from the same patients may give differing results [see 6, 17–19, 24, 25]. Our own studies have shown that the increase in leucocytes, including lymphocytes and eosinophils, occurs in relatively *mild* asthma (particularly of the intrinsic form) and that it is associated with the presence of irregularly-shaped lymphocytes together with 'activation' markers for both lymphocytes (ie CD25+ cells) and eosinophils (i.e. EG2+ cells) [6, 20, 21]. EG2 is a marker for the cleaved and secreted form of eosinophil cationic protein which can be found both within activated eosinophils and diffusely in the wall often in association with the reticular basement membrane. This highly charged protein and its associated major basic protein have been firmly implicated in damaging the airway mucosa and its surface epithelium. As with epithelial fragility, the increased activation of eosinophils also shows a significant relationship with AHR [21].

We have also shown up-regulation of gene expression for the lymphokine IL-5 in biopsies obtained from symptomatic atopic asthmatics [22] and this is also associated with AHR. The local release of interleukin 5 is likely to have important functional implications for the

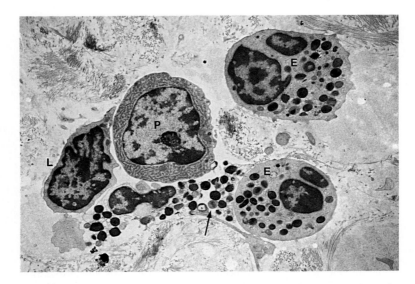

Fig. 4. Transmission electron micrograph showing a plasma cell (*P*), lymphocyte (*L*) and eosinophils (*E*) in the subepithelial layer of the bronchial mucosa of a subject with mild atopic asthma. One eosinophil appears to have degranulated (arrows). × 6750

eosinophil by stimulating differentiation of committed eosinophil precursors, favouring their selective adhesion, enhancing eosinophil survival in tissues and promoting their activation.

Studies of bronchoalveolar lavage have also shown increased numbers of eosinophils, recruitment of T-helper cells from blood with evidence of spontaneous mast cell and eosinophil degranulation [17, 23, 24]. The predominant profile of cytokine gene expression in cells examined in BAL includes IL3, IL4, IL5, GM-CSF and TNF alpha; constituting the so-called TH2 cytokine profile [35, 36]. These changes are not restricted to the bronchi. Alterations in the ratios of lymphocyte subsets, increases in the expression of markers of activation and increases of circulating basophil progenitors may also be detected in peripheral blood [25, 26].

Mast cells have long been considered to play a key role as a trigger cell in the immediate type sensitivitiy reaction to allergen challenge and their degranulation is associated with fatal acute severe asthma [27]. Their role in mild chronic ongoing disease is, however, less clear. Recent immunohistological [21, 28] and careful morphometric electron microscopic studies of bronchial biopsies in mild atopic asthma (Heard, Kay and Jeffery – unpublished) show no reductions of mast cell numbers nor of significant degranulation, suggesting their participation in mild, stable but chronic disease is minimal. Data from BAL show, however, that mast cells recovered from mild asthmatics spontaneously release their granular content more easily than normal [17]. Surprisingly, the precise role of the mast cell in asthma and in AHR is yet to be determined. The mast cell has also been shown to synthesize, store and secrete a number of proinflammatory cytokines: as such it may be an important initiator of the inflammatory cascade [37].

Bronchial vasculature, congestion and oedema

Dilatation of bronchial mucosal blood vessels with swollen endothelial cells, congestion and wall oedema are also features of fatal asthma (see Fig. 2). The increase in thickness of the bronchial wall in asthma is unlikely to be accounted for by the increase in muscle thickness and enlargement of submucosal gland size alone and may well be due to mucosal vessel dilatation, congestion and consequent wall oedema. James and colleagues have shown that the overall contribution of these components to airway wall thickening need only be relatively minor to have dramatic consequences in producing the pattern of airflow limitation seen in asthma [29]. Oedema of airway adventitial tissue may also cause uncoupling of airway wall from its surrounding alveolar attachments: this has been suggested to lead to loss of restraint as BSM contracts and hence to increased airway closure. Loss of elastic recoil in lung parenchyma may contribute similarly: however, studies carried out in the author's laboratory indicate that bronchial and alveolar wall elastic fibre content is normal in cases of fatal asthma [38].

Airway wall nerves

The topic of airway wall innervation is a large one [30, 31] and cannot be extensively discussed herein. There are, however, interesting data showing that in asthma there is an absence of (relaxant) vasoactive intestinal polypeptide (VIP) within nerve fibres [32] and an increase in the number of fibres containing substance P (stimulatory to bronchial smooth muscle) [33]: this contrasts markedly with the innervation of the 'control' lungs taken at resection from chronic smokers. If confirmed, this may provide yet another mechanism for predisposition to AHR.

Effects of treatment

We have obtained airway mucosal biopsies at fiberoptic bronchoscopy from each of 3 distinct airway levels of the left lung in 3 subject groups [19]. In eleven mild atopic asthmatics, (mean age = 29 yrs; FEV_1 % predicted normal 89–116%) we biopsied twice, once prior to 4 weeks treatment with either inhaled β-agonist (terbutaline: 250 mg, 2 puffs 4 times daily n = 5 subjects) or corticosteroid (budesonide: 200 µg, 1 puff twice daily n = 6 subjects) and again following this period to allow determination of the effects of treatment. The asthmatics, prior to treatment, had an increased cellular infiltrate compared with a healthy control group of 12 normal volunteers and a thickening of the reticular basement membrane. By transmission electron microscopy the infiltrate consisted of increases of eosinophils, eosinophil degranulation, and *decreases* of neutrophils. There was also an increase in the number of mast cells. Short-term treatment by the steroid (but not the β-agonist) did not alter the total numbers of infiltrating cells but did reduce mast cell and eosinophil numbers to normal and reduced the frequency with which foci of free eosinophil granules were found. Compared with healthy controls, a further group of asthmatics on long-term (average 3.5 years) corticosteroid treatment had a similarly low total cell infiltrate but still showed a significant thickening of reticular basement membrane ($p < 0.05$). Eosinophils were seen less frequently in this relatively more severe group than in the mild asthmatic group before treatment.

We have recently conducted a double-blind placebo-controlled study in which we examined bronchial biopsies for the effect of oral theophylline on the airway inflammatory late-phase reaction to allergen inhalation (n = 19 atopic asthmatics). Six weeks treatment with oral slow-release theophylline (200 mg, 12 hourly resulting in a mean serum concentration of 36.6 µmol/L) significantly reduced the allergen-induced eosinophil recruitment and their activation (as determined by EG2 immunostaining).

Our results indicate that decreases of eosinophil and mast cell numbers and reduction of eosinophil activation and degranulation, but not of the thickening of reticular basement membrane, are associated with clinically effective treatment.

Conclusion

These studies support the role of airway wall inflammation in bronchial asthma, confirm the involvement of eosinophils and their degranulation and implicate, further, T lymphocytes and interleukin 5 as controllers of the eosinophilic response. Several factors may contribute to apparent airways hyperresponsiveness including fragility of surface epithelium, thickening of the airway wall, uncoupling of airway wall and parenchyma or loss of lung elastic recoil. Thickening of the airway wall may be due to bronchial vessel dilatation and oedema, or enlargement of the mass of bronchial smooth muscle or mucus-secretory glands. Each of these changes may be induced by the inflammatory cells which infiltrate the airway wall in both mild and severe asthma: their predominant cytokine products can clearly influence the characteristic profile of cell phenotypes, their activation and the chronicity of the inflammatory response. In this regard, mast cells and lymphocytes may be important initiators and controllers, whilst activated eosinophils are key reactor cells whose increased presence is associated with increased airways responsiveness.

Acknowledgements

I am grateful to the National Asthma Research Campaign and Cystic Fibrosis Trust (UK) for their support, Mrs J. Billingham for secretarial and Mr A. Rogers for technical assistance.

References

1. Dunnill MS (1960) The pathology of asthma, with special reference to changes in the bronchial mucosa. J Clin Pathol 13: 27–33
2. Dunnill MS, Massarella GR, Anderson JA (1969) A comparison of the quantitative anatomy of the bronchi in normal subjects, in status asthmaticus, in chronic bronchitis, and in emphysema. Thorax 24: 176–179
3. Houston JC, De Navasquez S, Trounce JR (1953) A clinical and pathological study of fatal cases of status asthmaticus. Thorax 8: 207–213

4. Sobonya RE (1984) Quantitative structural alterations in long-standing allergic asthma. Am Rev Respir Dis 130: 289–292

5. Laitinen LA, Heino M, Laitinen A, Kava T, Haahtela T (1985) Damage of the airway epithelium and bronchial reactivity in patients with asthma. Am Rev Respir Dis 131: 599–606

6. Jeffery PK, Wardlaw A, Nelson FC, Collins JV, Kay AB (1989) Bronchial biopsies in asthma: an ultrastructural quantification study and correlation with hyperreactivity. Am Rev Respir Dis 140: 1745–1753

7. Callerame MD, Condemi MD, Bohrod MD, Vaughan JH (1971) Immunologic reactions of bronchial tissues in asthma. N Engl J Med 284: 459–464

8. McCarter JH, Vazquez JJ (1966) The bronchial basement membrane in asthma: immunohistochemical and ultrastructural observations. Arch Path 82: 328–335

9. Roche WR, Williams JH, Beasley R, Holgate ST (1989) Subepithelial fibrosis in the bronchi of asthmatics. Lancet i: 520–524

10. Crepea SB, Harman JW (1955) The pathology of bronchial asthma. I. The significance of membrane changes in asthmatic and non-allergic pulmonary disease. J Allergy 26: 453–460

11. Glynn AA, Michaels L (1960) Bronchial biopsy in chronic bronchitis and asthma. Thorax 15: 142–153

12. Heard BE, Hossain S (1973) Hyperplasia of bronchial muscle in asthma. J Pathol 110: 319–331

13. Azzawi M, Jeffery PK, Frew AJ, Johnston P, Kay AB (1989) Activated eosinophils in bronchi obtained at post-mortem from asthma deaths. J Clin Exp Allergy 19: 118

14. Gleich GJ, Motojima S, Frigas E, Kaphart GM, Fujisawa T, Kravis LP (1987) The eosinophilic leukocyte and the pathology of fatal bronchial asthma: evidence for pathologic heterogeneity. J Allergy Clin Immunol 80: 412–415

15. Beasley R, Roche W, Roberts JA, Holgate ST (1987) Cellular events in the bronchi in mild asthma and after bronchial provocation. Am Rev Respir Dis 139: 806–817

16. Cutz E, Levison H, Cooper DM (1978) Ultrastructure of airways in children with asthma. Histopathology 2: 407–421

17. Wardlaw AJ, Dunnett S, Gleich GJ, Collins JV, Kay AB (1988) Eosinophils and mast cells in bronchoalveolar lavage in mild asthma: relationship to bronchial hyperreactivity. Am Rev Respir Dis 137: 62–69

18. Adelroth E, Rosenhall L, Johansson SA, Linden M, Venge P (1990) Inflammatory cells and eosinophilic activity in asthmatics investigated by bronchoalveolar lavage: effects of antiasthmatic treatment with budesonide or terbutaline. Am Rev Respir Dis 142: 91–99

19. Jeffery PK, Godfrey RW, Adelroth E, Nelson F, Rogers A, Johansson S-A (1992) Effects of treatment on airway inflammation and thickening of reticular collagen in asthma: a quantitative light and electron microscopic study and correlation with BAL. Am Rev Respir Dis 145: 890–899

20. Azzawi M, Bradley B, Jeffery PK, Frew AJ, Wardlaw AJ, Assoufi B, Collins JV, Durham S, Kay AB (1990) Identification of activated T lymphocytes and

eosinophils in bronchial biopsies in stable atopic asthma. Am Rev Respir Dis 142: 1407–1413

21. Bradley BL, Azzawi M, Jacobson M, Assoufi B, Collins JV, Irani A-MA, Schwartz LB, Durham SR, Jeffery PK, Kay AB (1991) Eosinophils, T-lymphocytes, mast cells, neutrophils and macrophages in bronchial biopsies from atopic asthmatics: comparison with atopic non-asthma and relationship to bronchial hyperresponsiveness. J Allergy Clin Immunol 88: 661–674

22. Hamid Q, Azzawi M, Ying S, Mogbel R, Rance AJ, Wardlaw AJ, Corrigan CJ, Durham SR, Jeffery PK, Kay AB (1991) IL-5 mRNA in bronchial biopsies from asthmatic subjects. J Clin Invest 87: 1541–1546

23. Metzger WJ, Zavala D, Richerson HB, Moseley P, Iwamota P, Monick M, Sjoerdsma K, Hunninghake GW (1987) Local allergen challenge and bronchoalveolar lavage of allergic asthmatic lungs: description of the model and local airway inflammation. Am Rev Respir Dis 135: 433–440

24. Gerblich AA, Campbell AE, Schuyler MR (1984) Changes in T-lymphocyte subpopulations after antigenic bronchial provocation in asthmatics. N Engl J Med 310: 1349–1352

25. Denburg JA, Telizyn S, Belda A, Dolovich J, Bienenstock J (1985) Increased numbers of circulating basophil progenitors in atopic patients. J Allergy Clin Immunol 76: 466–472

26. Corrigan CJ, Hartnel A, Kay AB (1988) T-lymphocyte activation in acute severe asthma. Lancet 1: 1129–1132

27. Heard BE, Nunn AJ, Kay AB (1989) Mast cells in human lungs. J Pathol 157: 59–63

28. Djukanovic R, Wilson JW, Britten KM, Wilson SJ, Walls AF, Roche WR, Howarth PH, Holgate ST (1990) Quantitation of mast cells and eosinophils in the bronchial mucosa of symptomatic atopic asthmatics and healthy control subjects using immunohistochemistry. Am Rev Respir Dis 142: 863–871

29. James AL, Pare PD, Hogg JC (1989) The mechanics of airway narrowing in asthma. Am Rev Respir Dis 139: 242–246

30. Jeffery PK (1994) Innervation of the airway mucosa: structure, function and changes in airway disease. In: Goldie R (ed) Immunopharmacology of epithelial barriers. Academic Press, London, pp 85–118

31. Barnes PJ (1986) State of art: neural control of human airways in health and disease. Am Rev Respir Dis 134: 1289–1314

32. Ollerenshaw S, Jarvis D, Woolcock A, Sullivan C, Scheibner T (1989) Absence of immunoreactive vasoactive intestinal polypeptide in tissue from the lungs of patients with asthma. N Engl J Med 320: 1244–1248

33. Ollerenshaw SL, Jarvis D, Sullivan CE, Woolcock AJ (1991) Substance P immunoreactive nerves in airways from asthmatics and nonasthmatics. Eur Respir J 4: 673–682

34. Carrol N, Elliott A, Morton A, James A (1993) The structure of large and small airways in nonfatal and fatal asthma. Am Rev Respir Dis 147: 405–410

35. Robinson DS, Hamid Q, Ying S, Tsicopoulos A, Barkans J, Bentley AM, Corrigan C, Durham SR, Kay AB (1992) Predominant H2-like bronchial alveolar T-lymphocyte population in atopic asthma. N Engl J Med 326: 298–304

36. Broide DH, Lotz M, Cuomo AJ, Coburn DA, Federman EC, Wasserman SI
    (1992) Cytokines in symptomatic asthma airways. J Allergy Clin Immunol
    89 (5): 958–967
37. Bradding P, Feather IH, Howarth PH, Mueller R, Roberts JA, Britten K,
    Bews JPA, Hunt TC, Okayama Y, Heusser CH, Bullock GR, Church MK,
    Holgate ST (1992) Interleukin 4 is localized to and released by human mast
    cells. J Exp Med 176: 1381–1386
38. Lorimer S, Godfrey RWA, Majumdar S, Edelroth E, Johansson S-A, Jeffery
    PK (1992) Elastic fibre content of airway and lung parenchyma is not
    reduced in asthma. Eur Respir J 5: 65s (Abstract)
39. Sullivan P, Bekir S, Jaffar Z, Page C, Jeffery P, Costello J (1994) Anti-
    inflammatory effects of low-dose oral theophylline in atopic asthma. Lancet
    343: 1006–1008

Correspondence: P. K. Jeffery, MSc PhD, Lung Pathology, Department of
Histopathology, Royal Brompton National Heart & Lung Institute, GB-London
SW3 6NP.

Discussion

Moderator: Thank you Dr. Jeffery for these wonderful images explaining how things could happen, and for the additional verbal information which painted a very thorough picture of what we have to expect if we deal with the whole scope, from mild to the fatal asthma. Do you wish to ask any questions of Dr. Jeffery's presentation?

Dr. Poulter: One quick question. What do you think the pathological significance of the reticular thickening might be? And how do you relate that thickening to the observed increased permeability? Although you very clearly showed that cells could go through it, you might expect an overal decrease in permeability.

Dr Jeffery: I think there's insufficient work in regard permeability on the reticular basement membrane to make a comment. And certainly we haven't worked on that. I'm amazed at how easily the inflammatory cells seem to permeate it. I do however feel that it is permeable to macromolecules. I think it's not so much the thicknes of it but its' sugar composition and the binding of other molecules charged or not, such as heparin sulfate, which is going to control permeability rather than its structure per se. So I guess I can't answer that question, a lot more work has to be done in that regard. I can only speculate, in regard to its role in bronchial hyperresponsiveness. If you look at its thickness and to bronchial hyperresponsiveness, it doesn't show a statistically significant association. But I could offer just one speculation as to how its thickening might have an effect on bronchoconstriction, by changing the direction or the resolution of the force resulting from bronchial smooth muscle contraction. The increase in the reticular layer could increase the rigidity of the bronchial tube. If we go back to some very early anatomical studies by Miller and colleagues in the 30's, bronchial smooth muscle in fact doesn't encircle the airway, but rather winds its way down the airways in a helical fashion: In fact, there are two opposing helices (i.e. a geodesic pattern). So the effect of bronchial smooth muscle contraction is not only to squeeze the bronchial tube but to shorten it. If we speculate now that the thickened basement membrane retards the shortening of the airway, you could imagine that a lot more of that force, in its resolution, will go towards airway constriction.

Moderator: Do you know of somebody who has done some work on radiographic visualization of airway thickening?

Dr. Jeffery: Drs. Bousquet and Vignola have, and they see airway wall thickening.

Dr. Vignola: We did high-resolution computed tomography. And we performed millimetre thick slices. We were able to find irreversible alterations of the bronchial wall, as well as a heterogenity in the internal diameter of the bronchi in asthmatic patients as compared to normal subjects.

Moderator: Did you happen to go into the detail of how they behave under provocation? Is provocation visible on HR-CT? Is there variability in the morphological changes?

Dr. Vignola: Well, we first evaluated the internal diameter of the bronchi in non-stimulated asthmatic patients. And we then examined the modifications of the internal size of the bronchi after excercise challenge in patients who suffered from excercise-induced asthma. And we were able to find that after challenge, there really was a reduction of the internal size measured as number of pixels, reduction of the internal size of the bronchi after challenge, and the reduction of the internal diameter was not homogeneous in all the lobes of the lung. After inhalation of beta 2 agonists we were able to find that the internal size of the bronchi was increased. And in addition, some bronchi which were not seen at the baseline or after allergen challenge, were seen after the inhalation of beta 2 agonists, suggesting that at the baseline in the asthmatic condition there is even an obliteration of the airways.

Moderator: Thank you. Do you feel that this fibrosis or fibrous thickening of the reticular basement membranes is already a feature of mild asymptomatic asthma?

Dr. Jeffery: Yes, that's clearly the result form the work of our group and several others who have shown that in mild asthma you do get significant thickening. However, I have to say that the thickening is slight, an increase of about 2 to 4 microns in depth. But it is a homogenous thickening, and so the overall functional effect might be high. As you know, only small changes of airway diameter can have profound effects in airflow resistance to the power of four.

But I believe its mechanical features might be more important than the absolute amount of thickening that is observed. The sorts of changes that I showed were in fact in mild stable asthma (1, 2). If you go on to compare that thickening with what you see in an acute severe attack of asthma, the extra thickening that one sees in those fatal cases is actually very small in comparison to what you find even in mild asthma, so mild asthma or severe asthma, in regard to thickening of the reticular basement membrane, is of the same order of magnitude.

Moderator: Is there any reversible component? Did somebody recently do biopsies and examine them for reversibility of these changes?

Dr. Jeffery: Yes, several groups have now done that. In the study I have briefly presented to you, we looked at the effect of budesonide and compared it to beta 2 agonists (3). Terbutaline did not significantly reduce the inflammatory cell number, nor the reticular thickening we observed. Budesonide (200 µg twice daily) was certainly with effect, on reducing the inflammatory cell numbers, e.g. mast cell numbers, eosinophils, and eosinophil degranulation were all reduced after 4 weeks treatment with low doses of this steroid. But there was no significant reduction of the thickening of the reticular basement membrane over that 4-week period. However Dr. Saetta and Prof. Fabbri have looked at occupational asthma, following removal of exposure to the occupational agent (4). They do observe a return towards the normal in the thickness of the reticular layer. Prof. Davies and colleagues have also recently shown an effect of steroid (BDP) on reversibility after 5 months of treatment (5). So I think if you take these results all together, we're not yet sure, our results show that there was not a change within a 4-week period but maybe this is too short and longer term treatment will be effective.

Moderator: Is it too far fetched, that one could compare this mechanism of fibrosis with collagen related diseases, and secondly, might there be any effect of non-steroidal therapy? I recall statements (that really lack the proof) that these agents have been used in the treatment of chronic bronchitis (i.e. mucus-hypersecretion). Could you imagine that such a process could be influenced by anti-inflammatory drugs?

Dr. Jeffery: Probably so, I think there's a lot in the way of similarities between asthma and other inflammatory conditions of the lung. I showed data comparing T-cells in asthma with cystic fibrosis. And I am sure we will see very soon a number of publications which will show that many of the changes we have seen in asthma are not unique to asthma. In regard the cytokines, we're also currently looking at bronchial biopsies in chronic bronchitis. We have also completed the study in early fibrosing alveolitis associated with scleroderma (FASSc) (6). And interestingly, if you determine the cytokine profile ind FASSc: IL4 is high, IL5 high, Interferon gamma moderately high, but IL2 is low. So there is much similarity to the atopic situation. The exception is interferon gamma, which is relatively high in fibrosing alveolitis. And of course in the early stages and by lavage, there is an eosinophilia associated with fibrosing alveolitis which actually is a good prognostic indicator. So I think there will probably be room for looking at therapies in terms of anti-cytokine, IL 4 and IL 5. There may be effects in fibrosing alveolitis, similarly there will be effects in asthma. I don't think I'd like to speculate beyond that.

References

1. Jeffery PK, Wardlaw A, Nelson FC, Collins JV, Kay AB (1989) Bronchial biopsies in asthma: an ultrastructural quantification study and correlation with hyperreactivity. Am Rev Respir Dis 140: 1745–1753
2. Jeffery PK (1992) Pathology of Asthma. Br Med Bull 48: 23–39
3. Jeffery PK, Godfrey RWA, Adelroth E, Nelson F, Rogers A, Johansson SA (1992) Effects of treatment on airway inflammation and thickening of reticular collagen in asthma: a quantitative light and electron microscopic study. Am Rev Respir Dis 145: 890–899
4. Saetta M, Maestrelli P, Di Stefano A, De Marzo N, Milani GF, Mapp CE, Fabbri LM (1992) Effect of cessation of exposure to toluene diisocyanate (TDI) on bronchial mucosa of subjects with TDI-induced asthma. Am Rev Respir Dis 145: 169–174

5. Trigg CJ, Manolitsas ND, Wang J, Calderon MA, McAulay A, Jordan SE, Herdman MJ, Jhalli N, Duddle JM, Hamilton SA, Devalia JL, Davies RJ (1994) Placebo-controlled immunopathologic study of four months of inhaled corticosteroids in asthma. Am J Respir Crit Care Med 150: 17–22
6. Hamid Q, Majumdar S, Sheppard MN, Corrin B, Black CM, Du Bois R, Jeffery PK (1993) Expression of IL4, IL5, INF gamma and IL2 mRNA in fibrosing alveolitis associated with systemic sclerosis (in press). Am Rev Respir Dis 147: A 479 (Abstract)

Structural changes in the bronchial mucosa before and after therapy

T. Haahtela, A. Laitinen, and L. A. Laitinen

Department of Allergic Diseases and Pulmonary Medicine,
Helsinki University Central Hospital, Department of Anatomy,
University of Helsinki, Finland

Summary

Bronchoscopy is an invasive measure to the patient and can be compared to other endoscopic procedures, e.g. gastroscopy and colonoscopy. If the patient is well informed and prepared in advance, and the local anesthesia succeeds, the procedure is easy both for the patient and examiner. The safety aspects must be taken into account [28]. There are patients with very sensitive pharyngeal reflexes, and it is unwise to go on with the bronchoscopy with such patients. Complications are very rare.

The bronchial epithelium is injured at early stages of asthma and influx of inflammatory cells is observed. It seems also obvious that signs of asthmatic inflammation can be found in bronchial specimens even before the development of increased bronchial responsiveness and airflow limitation. This may offer a possibility for early detection of patients at risk. In everyday practice, however, we need simple and noninvasive methods for to detect early inflammation. Measurement for inflammatory markes, e.g. eosinophilic cationic protein (ECP) in spontaneous or induced sputum may be of help [29]. We need a basic reference for validation of these new methods, and tissue samples obtained by fiberoptic bronchoscopy may serve as such. While evaluating treatment effects and anti-inflammatory potency of various drugs, bronchial specimens gives direct information from the disease site. Bronchoalveolar lavage is more of a reflection what is really going on in the mucosal tissue. However, interpretation of the bronchial specimens should be standardized better in order to accomplish the goal for a "golden reference".

Fiberoptic bronchoscopy should gain wider acceptance as a basic research tool as well as a diagnostic aid for asthma and asthma-like symptoms in specialist practice. It has opened a new window to the understanding of the pathophysiology of asthma and to monitor it's progress and treatment effects.

Zusammenfassung

Strukturelle Veränderungen in der Bronchialschleimhaut vor und nach Therapie. Die Histologie als ein traditionell wichtiges Werkzeug hat nicht viel zur Überwachung der Atemwegsentzündung und der strukturellen Veränderungen beigetragen, da die Gewinnung von geeigneten Biopsieproben von lebenden Asthmatikern nicht immer zulässig ist und daher nicht als eine generell empfehlenswerte Untersuchung für Asthmatiker gelten kann. Allerdings konnte durch die Einführung der fiberoptischen Techniken eine Verbreiterung der klinischen Anwendung erzielt werden, sodaß auch der Therapieeffekt kontrolliert werden konnte.

Wir haben eine Serie von 14 Patienten mit neu entdecktem Asthma untersucht und sie 4 Kontrollpersonen gegenübergestellt. Bei den Asthmatikern fand sich eine gesteigerte Zahl von Mastzellen, Eosinophilen, Lymphozyten, Makrophagen und Plasmazellen im Epithel oder in der Lamina propria im Vergleich zu den Kontrollen. Es bestand eine Becherzellhyperplasie und eine Verminderung der zilientragenden Zellen. Diese Untersuchung zeigte, daß in den asthmatischen Luftwegen, sogar bei Patienten mit neu entdeckter Erkrankung, alle Zeichen der entzündlichen Reaktion mit ausgeprägten strukturellen Veränderungen nachweisbar sind.

Die entzündlichen Veränderungen in der Bronchialmukosa und die Steigerung der bronchialen Reaktivität wurden bei einem asthmatischen Patienten während der Verschlechterung seiner Symptome untersucht. Durch wiederholte Biopsien konnte eine markante Steigerung in der Zahl der Eosinophilen im Epithel nachgewiesen werden. Die Präparate zeigten auch vermehrte Permeabilität, da nämlich Zellen während der Penetration des Gefäßendothels beobachtet wurden, begleitet von Hyperplasie der Becherzellen, Epithelmetaplasie und Anhäufung von Ödemflüssigkeit mit Proteinen und Makromolekülen. Nach der Behandlung mit einem inhalativen Kortikosteroid (Budesonide) durch 16 Wochen war das Epithel weitgehend wieder hergestellt, wenngleich es noch ziemlich brüchig erschien.

Der Effekt von inhalativem Budesonide und Terbutaline auf die klinische Symptomatik, die Lungenfunktion und die Atemwegsentzündung wurde bei 14 Patienten mit frisch entdecktem Asthma untersucht. Wieder konnte durch bronchiale Biopsien demonstriert werden, wie die Struktur des Epithels sich besserte und die Zahl der meisten Entzündungszellen abnahm. Die Zahl der intraepithelialen Nervendigungen nahm während der Kortikosteroidtherapie zu.

Inhalative Kortikosteroide haben eine breite Wirkung auf Reparationsvorgänge der entzündeten Bronchialschleimhaut. Sie vermindern die Gesamtzahl der Entzündungszellen. In unserer Studie vermochte Budesonide die Zahl der Eosinophilen und der Lymphozyten im Epithel und die Zahl der Lymphozyten und der Plasmazellen in der Lamina propria zu vermindern. Allerdings steigerte Budesonide die Zahl der Fibroblasten in der Lamina propria. Ein inhalatives β-2-Mimetikum, Terbutalin, hatte ebenfalls einen Effekt auf die Lymphozyten und die Plasmazellen. Das Verhältnis der zilientragenden und der Becherzellen, wie auch die Zahl der intraepithelialen Nerven konnte durch die Behandlung mit einem inhalativen Kortikosteroid gesteigert werden, während die Zahl der extrazellulären Matrixproteine, Tenascin und vielleicht Kollagen III und VII in der Basalmembran abnahmen. Außerdem konnten inhalative Kortikosteroide die Gefäßpermeabilität vermindern, was sich in einer Verminderung der Zahl der schmalen Klüfte in den postkapillären Venolen manifestierte.

In einer Untersuchung von Bronchialbiopsien zeigte sich, daß Nedocromil einen marginalen Effekt auf die Zahl der Eosinophilen im Epithel im Vergleich zu den β-2-Agonisten hatte.

Introduction

Direct monitoring of the airway inflammation and structural changes in asthma has not been much used because of the difficulties in obtaining specimens of bronchial mucosa in living asthmatics and because of lack of motivation. Taking bronchial biopsies has several limitations and methodological problems and is not a procedure for asthmatic subjects by and large. However, along with the development of the fiberoptic techniques, this method has been adapted for wider clinical use both for diagnostic purposes and for evaluation of the effect of various therapies. There are several important advantages of the classically important tool, histology:

1. Bronchial biopsies offer direct information from the pathological changes in the airways.
2. Tissue samples represent material from actual localization of the disease.
3. Tissue samples may be processed for several different methods (light-, transmission- and scanning electron microscopy, immuno-histochemistry, biochemistry, tissue cultures etc.
4. Mucosal specimens provide quantitative information on tissue pathology at cellular and subgross histology level.

There are multiple scientific indications for investigational bronchoscopy, and it is also a frame of reference for other methods more readily appilicable to routine investigations especially in general practice. This article is concentrated to the morphological findings, which can be assessed by bronchial mucosal tissue sampling before and after various asthma therapies. At the moment, however, only a few studies of this kind have been made. In addition we discuss some of the methodological aspects of taking and interpreting bronchial biopsies.

Obtaining the specimens

Obtaining representative specimens from the human airways is the main problem of morphological studies. Traditionally, asthma patients have been bronchoscopied mainly for the diagnosis of infiltrative diseases or tumors. Only in the past few years has the asthmatic process itself been the object and indication for the procedure.

For our first studies we used the rigid-tube bronchoscope in order to obtain large enough specimens from the mucosa. This method results in a high yield of representative airway specimens. In one study, proper airway morphology with epithelium was obtained from 96% of the biopsies in the asthmatic subjects and from 75% in the control subjects [1]. This is in sharp contrast to the result with fiberoptic technique, where two thirds of 120 biopsies had to be discarded since they showed extensively damaged epithelium or no epithelium at all [2]. In our series of 26 biopsies the specimens showed epithelium with a mean length of 710 μm (300 to 1160 μm). The lamina propria under the corresponding epithelium and basement membrane had a mean depth of 190 μm (from 100 to 250 μm) [1].

For practical reasons, however, we use now mostly the fiberoptic technique, although good samples are not always obtained and they are essentially smaller than with rigid-tube. The major advantage of fiberoscopy is, that there is no need to hospitalize and sedate the patient. Before the procedure, the chest is x-rayed and lung volumes are measured. EKG is recorded and blood clotting and bleeding times are defined. An hour before the bronchoscopy the patient is given petidin and atropin im., and 30 min before 8 ml nebulized lidocaine to inhale (20 mg/ml). An extra dose of a bronchodilator is sometimes given.

The examination is performed while the patient is sitting opposite the examiner. If needed, he is given an additional local anesthesia under visual control by slowly dropping lidocain into the larynx and trachea. The fiberoscope, the tip of which is lubricated by lidocain gel, is introduced through the mouth to the trachea. This may cause irritation and cough, which usually disappears after intubation. The procedure is made as gently as possible in order to avoid irritation and cough. The sites of biopsies are not touched with the fiberoscope before taking the actual specimens. Usually 4–8 specimens are taken from standard airway levels, and they are immediately submerged in the fixation solution (10% formaldehyde, ethanol or 3% glutaraldehyde). The bronchoscopic examination takes 15–20 min on an average. The patient is observed for at least 2 hours before leaving the hospital. The experiences are good and no severe complications have occurred.

The major problems ofthe methos are:

1. Little reference material is available both from subjects and disease groups.
2. Specimens are relatively small and need validation in correlation to the airways as a whole.

3. The procedure is invasive, requires systemic and topical medication and induces a degree of trauma in the airways.
4. The procedure cannot be performed frequently and it is difficult to perform in normal subjects.
5. Special skill is required to properly master the technique.
6. The procedure as a whole is fairly expensive.

Interpreting the results with light- or electron microscopy

In order to get valuable information, the bronchial biopsy specimens should be taken, processed and analyzed in a standardized fashion. No joint reserach has been done to accomplish this goal and various laboratories use their own standards and principles to express the results.

Light microscopy is much less laborious and time consuming than electronmicroscopy, which demands considerable skill and experience. Using light microscopy, also large areas of the biopsy can be studied. The recognition of different kinds of inflammatory and other mobile cells is usually based on their different staining properties. Because of the limitations in the magnification with this method, little information about the fine structure of the individual cells and structures not visible in small magnifications, such as nerves, can be obtained. There is a real clinical need for a validated and at least a semi-quantitative method to express the findings in light microscopy: loss of ciliated cells, goblet cell hyperplasia, the relation of ciliated cells to goblet cells, metaplasia, shedding of the epithelium, amount of the most important inflammatory cells, accumulation of oedema fluid, thickness of basement membrane etc.

In electron microscopy, the surface area of the specimen is much more restricted than in light microscopy. In order to avoid confusion between different morphological studies, the methods used should be described in a precise way, and if possible the quality, as well as the size of individual specimens, should be visualized in a study produced by quantitative electron microscopy [3]. Only with electron microscopy the fine structures of individual cells and structures can be evaluated. In disease states, inflammatory cells can also show changes in morphology such as degranulation, which cannot be judged with light microscopy.

Morphological changes in bronchial mucosa before and after treatment

Asthmatic patients compared to normal subjects

Already in 1960s epithelial shedding and influx of eosinophils into the airway mucosa were associated with asthma [4, 5].

In 1980s we took fresh biopsies from 8 asthmatics (2 mild, 3 moderate and 3 severe) and showed that they all had more or less epithelial destruction and shedding at the three airway levels studied [6]. This was in contrast with the finding in a control subject, who showed in 3 specimens perfectly intact epithelium. The ciliated cells were the most affected. Intraepithelial nerves and mast cells were indentified. Because the study was focused on the epithelium and not on the subepithelial tissue, other inflammatory cells besides mast cells and some leukocytes were not detected. The study revealed profound pathological changes in the asthmatic epithelium. Perhaps the most important observation was, that marked changes can also exist in patients with newly detected asthma, who show clinically and functionally a mild disease.

Ollerenshaw and Woolcock [7] showed intact epithelium along 84% of the basement membrane in healthy control subjects and in 56% in mild to moderate asthmatics. However, Lozewicz et al. [8] took biopsy specimens of the bronchial mucus membrane from 11 patients with mild atopic asthma and suggested that airway hyperresponsiveness may occur without apparent changes in the structure of the bronchial epithelium.

Since then the asthmatic inflammation has been characterized in more detail. For example, Djukanovic et al. [9] showed in the bronchial mucosa of symptomatic atopic asthmatics a significant increase of mast cells and eosinophils compared with healthy controls. Intraepithelial lymphocytes have been shown to form the major proportion of intraepithelial migratory cells [10], and Azzawi et al. [11] demonstrated that the highest number of CD45-, CD3-, CD4-, and CD8-positive cells are found in asthmatics when compared with nonasthmatic atopic or healthy control subjects. Also the number of macrophages has been clearly increased in asthmatic bronchi [12].

We studied a series of 14 patients with newly detected asthma and

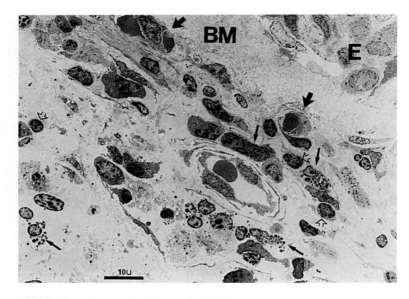

Fig. 1. Transmission electron micrograph. The airway specimen is from a newly defined asthmatic patient. Beneath the airway epithelium (*E*) is seen a thickened basement membrane (*BM*). Under the BM in the stroma are seen numerous capillaries (arrowheads) and many inflammatory cells such as eosinophils (closed arrows), lymphocytes (open arrows) and plasma cells reflecting chronic inflammation of the airways. Original magnification × 1.300. Bar = 10 μm. From reference [1]

compared to 4 control subjects to further evaluate the morphological differences [1]. There was an increase in the numbers of mast cells, eosinophils, lymphocytes, macrophages and plasma cells in the epithelium or lamina propria on patients with asthma (Fig. 1) as compared to the control subjects. The airway epithelium showed a variety of different morphological forms, whereas in the control subjects it was of the pseudostratified ciliated type. Goblet cell hyperplasia occurred with or without ciliated cells and the number of ciliated cells was decreased in relation to goblet or all epihelial cells. The result showed that in the asthmatic airways, even in patients with mild and new asthma, there are signs of general inflammatory response already with structural changes.

Disodium chromoglycate

Although chromoglycate has been used for decades in the treatment of asthma, there are no controlled bronchial biopsy studies of its effect on mucosal structure or inflammation. Backman et al. [26] took bronchial biopsies on 17 children aged 6–17 years who had been taken chromogycate about two years. Nine were described as normal, six as showing mild inflammation and two as moderate inflammation. The type of inflammation was not characterized in detail.

Nedocromil sodium

Trigg et al. [27] made a placebo controlled bronchial biopsy study to investigate the effect of 4 months treatment with nedocromil sodium or salbutamol in 38 mild adult asthmatics. Seven of 9 subjects receiving nedocromil sodium had reduced number of eosinophils after treatment while 10 of 11 subjects receiving salbutamol had increased levels. The number of eosinophils did not change in the placebo group. No treatment differences were detected on mast cell numbers.

Conclusions

Bronchoscopy is an invasive measure to the patient and can be compared to other endoscopic procedures, e.g. gastroscopy and colonoscopy. If the patient is well informed and prepared in advance, and the local anesthesia succeeds, the procedure is easy both for the patient and examiner. The safety aspects must be taken into account [28]. There are patients with very sensitive pharyngeal reflexes, and it is unwise to go on with the bronchoscopy with such patients. Complications are very rare.

The bronchial epithelium is injured at early stages of asthma and influx of inflammatory cells is observed. It seems also obvious that signs of asthmatic inflammation can be found in bronchial specimens even before the development of increased bronchial responsiveness and airflow limitation. This may offer a possibility for early detection of patients at risk. In everyday practice, however, we need simple and noninvasive methods for to detect early inflammation. Measurement for inflammatory markes, e.g. eosinophilic cationic protein (ECP) in spontaneous or induced sputum may be of help [29]. We need a basic reference for validation of these new methods, and tissue samples obtained by fiberoptic bronchoscopy may serve as such. While evaluating treatment effects and anti-inflammatory potency of various drugs, bronchial specimens gives direct information from the disease site. Bronchoalveolar lavage is more of a reflection what is really going on in the mucosal tissue. However, interpretation of the bronchial specimens should be standardized better in order to accomplish the goal for a "golden reference".

Fiberoptic bronchoscopy should gain wider acceptance as a basic research tool as well as a diagnostic aid for asthma and asthma-like symptoms in specialist practice. It has opened a new window to the understanding of the pathophysiology of asthma and to monitor it's progress and treatment effects.

References

1. Laitinen LA, Laitinen A, Haahtela T (1993) Airway mucosal inflammation even in patients with newly diagnosed asthma. Am Rev Respir Dis 147: 697–704
2. Söderberg M, Hellström S, Sandström T, Lundgren R, Bergh A (1990) Structural characterization of bronchial mucosal biopsies from healthy volunteers; a light and electron microscopical study. Eur Respir J 3: 261–266
3. Haahtela T, Laitinen A, Laitinen LA (1993) Using biopsies in the monitoring of inflammation in asthmatic patients. Allergy 8: 65–69
4. Glynn AA, Michaels L (1960) Bronchial biopsy in chronic bronchitis and asthma. Thorax 15: 142–153
5. Dunnill MS, Massarella GR, Anderson JA (1969) A comparison of the quantitative anatomy of the bronchi in normal subjects, in status asthmaticus, in chronic bronchitis and in emphysema. Thorax 24: 176–179
6. Laitinen LA, Heino M, Laitinen A, Kava T, Haahtela T (1985) Damage of the airway epithelium and bronchial reactivity in patients with asthma. Am Rev Respir Dis 131: 599–606
7. Ollerenshaw SL, Woolcock AT (1992) Characteristics of the inflammation in biopsies from large airways of subjects with asthma and subjects with chronic airflow limitation. Am Rev Respir Dis 145: 922–927
8. Lozewicz S, Wells C, Gomez E, Ferguson H, Richman P, Devalia J, Davies RJ (1990) Morphological integrity of the bronchial epithelium in mild asthma. Thorax 43: 1215
9. Djukanovic R, Wilson JW, Britten KM et al (1990) Quantification of mast cells and eosinophils in the bronchial mucosa of symptomatic atopic asthma-

tics and healthy control subjects using immunohistochemistry. Am Rev Respir Dis 142: 863–871

10. Jeffrey PK, Wardlaw AJ, Nelson FC, Collins JV, Kay AB (1989) Bronchial biopsies in asthma. An ultrastructural quantitative study and correlation with hyperreactivity. Am Rev Respir Dis 140: 1745–1753

11. Azzawi M, Bradley B, Jeffery PK et al (1990) Identification of activated T lymphocytes and eosinophils in bronchial biopsies in stable atopic asthma. Am Rev Respir Dis 142: 1407–1413

12. Poston RN, Chanez P, Lacoste JY et al (1992) Immunohistochemical characterization of the cellular infiltration in asthmatic bronchi. Am Rev Respir Dis 145: 918–921

13. Laitinen LA, Laitinen A, Heino M, Haahtela T (1992) Eosinophilic airway inflammation during exacerbation of asthma and its treatment with inhaled corticosteroid. Am Rev Respir Dis 143: 423–427

14. Laitinen LA, Laitinen A, Haahtela T (1992) A comparative study of the effects of an inhaled corticosteroid, budesonide, and a B2-agonist, terbutaline, on airway inflammation in newly diagnosed asthma: a randomized, double-blind, parallel-group controlled trial. J Allergy Clin Immunol 90: 32–42

15. Haahtela T, Järvinen M, Kava T et al (1991) Comparison of a B2-agonist, terbutaline, with an inhaled corticosteroid, budesonide, in newly detected asthma. N Engl J Med 325: 388–392

16. Haahtela T, Järvinen M, Kava T et al (1994) Effects of reducing or discontinuing inhaled budesonide in patients with mild asthma. N Engl J Med 331: 700–705

17. Sullivan O, Bekir S, Jaffar Z et al (1994) Anti-inflammatory effects of low-dose oral theophylline in atopic asthma. Lancet 343: 1006–1008

18. Laursen LC, Taudorf E, Borgeskov S et al (1988) Fiberoptic bronchoscopy and bronchial mucosal biopsies in asthmatics undergoing long-term high-dose budesonide aerosol treatment. Allergy 43: 284–288

19. Jeffrey PK, Godfrey RW, Ädelroth E et al (1992) Effects of treatment on airway inflammation and thickening of basement membrane reticular collagen in asthma. Am Rev Respir Dis 145: 890–899

20. Burke C, Power CK, Norris A et al (1992) Lung function and immunopathological changes after inhaled corticosteroid therapy in asthma. Eur Respir J 5/1: 73–79

21. Lundgren R, Söderberg M, Horstedt P, Stenling R (1988) Morphological studies of bronchial mucosal biopsies from asthmatics before and ten years of treatment with inhaled steroids. Eur J Respir Dis 1: 883–889

22. Djukanovic R, Wilson JW, Britten KM et al (1992) Effect of an inhaled corticosteroid on airway inflammation and symptoms in asthma. Am Rev Respir Dis 145: 669–674

23. Trigg CJ, Manolitsas ND, Wang J et al (1994) Placebo-controlled immuno-pathologic study of four months of inhaled corticosteroids in asthma. Am J Respir Crit Care Med 150: 17–22

24. Sousa AR, Poston RN, Lane SJ, Nakhosteen JA, Lee TH (1993) Detection of GM-CSF in asthmatic bronchial epithelium and decrease by inhaled corticosteroids. Am Rev Respir Dis 147: 1557–1561

Correspondence: Dr. Tari Haahtela, Department of Allergic Diseases, Helsinki University Central Hospital, Meilahdentie 2, FIN-00250 Helsinki, Finland.

Discussion

Moderator: Thank you Dr. Haahtela for your beautiful presentation. And your paper is open for discussion.

Dr. Jeffery: That was an excellent presentation. And I would agree with you in regard the difference between terbutaline and budesonide, although ours was a much shorter study, and I compliment you on these long-term studies which are very much required. One thing we also did in a separate group, we biopsied once, in a group of patients given steroid for many years, on average three and a half years, but some were treated for as long as 10 years. We observed that there was definitely a decrease in the number of eosinophils. But one of the things that struck us was that there were still some eosinophils remaining (Ref. 3, see p. 87). I noticed that on one of your slides that there were some eosinophils left in the epithelium after treatment. My question to you is: did you look beneath the epithelium? Did you still find eosinophils in your corticosteroid-treated group? And if so, how do you interpret that? My feeling is that, as you've already implied, corticosteroid is not curing our asthma, it is rather like making a dam to hold back the eosinophils. As soon as you remove the corticosteroid, the eosinophils all flow back again. So we're not effectively treating the asthma, we're just treating the symptoms.

Dr. Haahtela: I partly agree with you. I also think that steroids are symptomatic drugs, but they are closer to the pathogenetic root than beta 2 agonists. But I also disagree a little bit. If we really tackle the inflammation early enough, and the patient is treated long enough, it may influence the outcome. It may influence also the outcome of hyperresponsiveness. We haven't yet published the 3-year results of our long term clinical study, but it seems that after 2 years'treatment with 1.2 mg of budesonide we can stop the treatment altogether in about one-third or half of patients. We gave them placebo and they didn't exacerbate anymore. This is very promising and gives us some optimism. So, I'm a bit hesitant to say that each intervention with inhaled corticosteroids is just symptomatic.

Dr. Lane: Were any of your patients resistant to corticosteroids clinically? And if so, was this reflected in what you found in biopsies?

Dr. Haahtela: In these studies we haven't got any patient with newly detected asthma who had been corticosteroid-resistant. All of these patients who were allocated to receive budesonide responded clinically. Probably there is a genuine corticosteroid resistance but I haven't seen it in these newly detected patients.

Dr. Lundgren: I was wondering in terms of whether you've seen effects or not, with the treatment of steroids, and I guess that's the worry with all bronchial studies because it seems that the site of real obstruction in asthma, is much more peripheral than what you can ever get in the biopsy. Isn't that a worry in the interpretation of your data?

Dr. Haahtela: Yes, it worried us. Nevertheless, we have done it for years and taken biopsies often from 3 airway levels, the third being inside of the right lower lobe, so that is not under the visual control. Sometimes it proved to be, in fact, a transbronchial biopsy. Thus the third biopsy came rather from the periphery. The results were, however, very much the same. If you take the specimen form the central airways, at the opening of the right upper bronchus, or inside the right lower bronchus, the result was more or less the same. Mucosal inflammation in asthma seems to be a general response.

Dr. Jeffery: We have examined the left lung with Dr. Ellinor Ädelroth (Ref. 3, see p. 87). While we took several biopsies from deep down in the basal segments, they showed similar results to those taken more proximally. We also did a study at the Brompton where we wanted to examine the contribution of each component to the total variation in counts. The biggest contributions to the coefficient of variation was that observed between subjects. 80% of the variation was accounted for in regard variation between subjects. The airway level variation was extremely small. In fact, the variation within one biopsy was greater than the variation between airway levels. The results indicated that we had to have at least 16 patients in a group and we needed to sample our biopsy repeatedly, but that we didn't have to be too concerned about the variation between airway levels. Independently, we feel that these biopsies albeit small can reflect what is going on throughout the airways.

Dr. Poulter: I'm not a clinician, but if I were a clinician and I gave an asthmatic inhaled corticosteroids, I would probably expect to see some clinical improvement before three months of therapy. Now if you accept that, do you interpret the modest changes that you and indeed other groups see at three months to mean that you can have clinical improvement without any change to the inflammatory reaction? Or do you think these changes occur much, much earlier? And you only happen to be seeing them at 3 months, because that's when you're taking the biopsy.

Dr. Haahtela: Clinically we saw an effect in 3 to 4 days. And I'm quite convinced that if we would just take new biopsies after 3 days, we would have seen the effect also at the cellular level.

N.N.: It would be very very interesting to take serial biopsies from the first day. From when you start the treatment, just the next morning there would be a difference I think. But would you like to be the patient for the serial biopsies?

Moderator: Would you be able to say that what you see were activated cells, and after such a short period of time, maybe the cells haven't had time to disappear.

Dr. Haahtela: In fact we did also a study on patients who inhaled leukotriene E4 and eosinophilic inflammation was observed in the bronchial wall 4 hours after the inhalation. The eosinophils were activated, based on the outlook of the granules.

Dr. Poulter: How about the lymphocytes? Do they tell you, "I am activated"?

Dr. Haahtela: No, but from mast cells and eosinophils you can tell something about their stage of activation by judging degranulation.

Dr. D'Alonzo: Is it always necessary to do biopsies? What role does BAL play in this picture? How can we learn a substantial amount of information say in performing therapeutic interventional studies using BAL as a substitute for biopsy? And if so, does the cellular picture have a reasonable relationship with bronchial hyperreactivity? Really looking at the technique is an important parameter of doing therapeutic interventional work.

Dr. Haahtela: Yes, of course a lot of studies have been done just with BAL. But we haven't used the bronchoalveolar lavage so much becauce we know that the small biopsies won't harm the patient, and in many cases BAL is even more strenuous to the patient. It takes more time, and if you put lots of saline into the patient's lungs, more side effects occur. The lavage gives kind of an indirect information about what is happening in the mucosa, the biopsy gives more direct information.

Dr. D'Alonzo: The advantage of BAL would be that it's less labour-intensive from the standpoint of morphological definition. If you have to resort to electrone microscopy to complete your definition, it takes a great deal more time and expense.

Dr. Haahtela: That's right.

Dr. Poulter: If I can have a comment, we did a study where we actually compared and matched lavage and biopsy from the same individuals. Our conclusion was that the lavage does not reflect quantitatively what is going on in the tissues at all. And as far as I am aware, of about 10 papers that investigated lavage, only 1 paper suggested that it did reflect the situation in the mucosa. Although it may be less labor-intensive afterwards, it's also our experience that lavage is not as well-tolerated by the asthmatic patients as biopsy. And you run an increased risk of bronchial constriction when your're bronchoscoping these people, if you do a lavage. Whereas we've had no problems whatsoever taking biopsies. So I think it's a very valuable tool for collecting samples of airway cells and I think there are several studies including those of John Costello, where people have artificially challenged through the bronchoscope and then lavaged, to look at acute events following challenge, and there may be the real value of lavage. But in terms of simply sampling the lung of the asthmatic, lavage in my view is in no way a substitute for the biopsy.

Dr. Jeffery: Yes, I could just endorse these comments. We did a similar study and in the asthmatic we found no correlation between the number of each cell phenotype in the lavage and that present in the bronchial mucosa. On the other hand we did find a correlation in normal healthy individuals, so it does underline again the variability that is seen in asthma. Interestingly though, Doug Robinson in his studies found a better correlation if you look at the mRNA for cytokines between BAL and mucosa. In contrast to studies with BAL I do feel that Freddy Hargreave's initiative in regard looking at sputum as a possible indicator of mucosal changes is a valuable one: it needs validation by comparisons of biopsy with sputum, in the same subjects. I think sputum might be a less invasive way forward.

Dr. Haahtela: I very much agree, because we have also studied sputum. For example, inflammatory markers like ECP show up beauti-fully in the sputum, while they are not changing at all in the serum.

Dr. Jeffery: Spontaneous or induced sputum?

Dr. Haahtela: Well, we have used induced sputum but now moving on to validate with spontaneous sputum. In general practice there is an urgent need for new and simple diagnostic tools to improve detection and follow-up of mucosal inflammation.

Mucus production in the lower airways – therapeutic implications

J. D. Lundgren

Department of Infectious Diseases (7722), University State Hospital,
Tagensvej, Copenhagen N, Denmark

Summary

Increased mucus production in the lower airways during exacerbation of bronchial asthma may contribute to airway obstruction. A variety of pathophysiological mechanisms may cause the increase in mucus production including imbalances in the autonomic nervous system including release of neuropeptides and activation of the cholinergic nervous system, increased release of secretagogues from epithelial cells and release of secretagogues from accumulated inflammatory cells, e.g. basophilic, eosinophilic and neutrophilic granulocytes.

Clinical trials documenting the efficacy of therapeutic agents in controlling the enhanced mucus production are not available. However several of the regimes in current use may modulate mucus secretion and/or transport. Rational approaches to prevent the increased production of mucus in asthma includes:

1. manipulation of the nervous system,
2. use of antiinflammatory agents, and
3. increase clearence.

Of licensed drugs, glucocorticoids, beta-adrenergic agonists, cholinergic antagonists and possible theophylline may be beneficial in the management of mucus hypersecretion in asthma.

Zusammenfassung

Schleimsekretion und deren pharmakologische Beeinflußbarkeit.
Die gesteigerte Schleimproduktion in den tiefen Luftwegen während einer Exazerbation des Asthma bronchiale kann maßgeblich zur Obstruktion beitragen. Eine Reihe von pathophysiologischen Mechanismen kann die Schleimproduktion steigern:

1. Balancestörungen im autonomen Nervensystem, einschließlich Freisetzung von Neuropeptiden und Aktivierung von cholinergischen Nerven,
2. gesteigerte Freisetzung von Sekretagoga aus Epithelzellen und aktivierten Mastzellen,
3. Freisetzung von Sekretagoga aus Ansammlungen entzündlicher Zellen, z.B. eosinophilen und neutrophilen Granulozyten.

Die Aktivierung des beta-adrenergischen Rezeptors oder eine intrazelluläre Steigerung des cAMP beeinflußt wohl nicht direkt die Schleimproduktion, ist aber für den verbesserten Abtransport von Schleim aus den Luftwegen verantwortlich. Die Sekretagoga, welche von ortsansässigen und hinzugekommenen Zellen freigesetzt werden, bestehen aus Eikosanoiden (Prostaglandine und Peptyl-Leukotriene), plättchenaktivierendem Faktor (PAF), eosinophilem kationischen Protein (ECP) und neutrophiler Elastase.

Klinische Prüfungen, welche die Wirksamkeit von therapeutischen Stoffen bei der Behandlung von gesteigerter Mucusproduktion dokumentieren, sind bislang nicht verfügbar. Ein vernünftiger Zugang zur Verminderung der gesteigerten Schleimproduktion beim Asthma schließt folgende Schritte ein:

1. Beeinflussung der nervösen Steuerung
2. Anwendung von entzündungshemmenden Medikamenten
3. Expektorantien.

Von den registrierten Medikamenten sind die Anticholinergika und die Glukokortikoide wahrscheinlich besonders nützlich, es werden aber ständig neue Medikamente entwickelt.

Introduction

An increasing appreciation of the role inflammation in the acute, and late phase asthmatic responses has become evident, and control of the inflammatory reaction is now a primary focus in the management of asthmatics [1]. There are several effector cells in the airways that may interact with the inflammatory reaction leading to alterations of the homeostasis of the airways and in the development of an asthmatic attack. These effector cells cause: i) activation of the smooth muscle cells leading to bronchoconstriction, ii) leakage from the capillaries of the submucosa leading to oedema formation, iii) destruction of the surface epithelium leading to exposure of sensory nerves, a more readily access of allergens and other irritants to the submucosal area, and decreasing the mucociliary clearence rate, and iv) activation of the mucus producing cells leading to increased mucus production [2]. This review will focus on current knowledge of the regulation of mucus production in the lower airways of asthmatics.

There are several lines of evidence suggesting that increased mucus production may contribute in the pathophysiology of asthma. Twenty to eighty percent of asthmatics may report productive cough during an attack [3]. As healthy individuals do not have expectoration, this finding may be taken as evidence that more than a normal volume of mucus is produced in the lower airways of asthmatics. Bronchi of fatal asthmatic cases may contain impacted mucus with numerous eosinophil granulocytes [4]. Furthermore, as will be discussed below, mediators released during an astmatic attack increase release of mucus from isolated airways (Table 1), and furthermore currently available treatment may inhibit the production of mucus in the airways.

Bronchoobstruction as measured by decrease in peakflow or forced expiratory volume within the first second of an exhalation (FEV_1) is the hallmark to evaluate the severity of asthma. To which extend mucus in the airways affects these measurements are not clearly understood. However, theoretically – at a given pressure difference between the surrounding air and the alveoli – if the airway diameter decreases by 10% the air-flow rate decreases by 25–35% [5]. Furthermore, James et al. [6] estimated from a computer model that the airway resistance may increase 100 times in asthmatic airways after maximal muscle shortening, whereas resistance may only increase 7 times in non-asthmatic

Table 1. Identified mucus secretagogues in human airways in vitro and/or in vivo

Source	Name
Parasympathetic nerves	Acetylcholine, vasoactive intestinal peptide
Sympathetic nerves	Norepinephrine (only via activation of alpha-adrenergic receptors)
Sensory nerves	Substance P, gastrin releasing peptide
Inflammatory cells	All: Eicosanoids, platelet activating factor, oxygen free radicals
Mast cells	Histamine
Eosinophilic granulocytes	Eosinophilic cationic protein
Neutrophilic granulocytes	Neutrophil elastase
Monocytes/macrophages	Monocyte/macrophage mucous secretagogues (MMS)[a]

[a] Secretegagoues ranging in size from 2–67 kDalton has been reported [34]

airways. In the study by James et al. [6], only increase in thickness of the asthmatic airway mucosa was assessed, but if mucus accumulation in the airway lumen had also been studied this would presumably have widened these differences even more. In any event, total occlusion of the airways by impacted mucus, as observed in fatal cases, will result in formation of atelectases and consequently in ventilation-perfusion mismatch.

Thus, mucus hyperproduction occurs during an asthmatic attack and this dys-regulation may contribute to the bronchoobstruction.

Constituents of mucus

The goblet cells and the submucosal gland mucus cells are the primary manufacturer of the acidic macromolecules constructed as glycoproteins (Table 2). The chemical structure of these glycoproteins are the primary reason for the visco-elastic properties of mucus [7]. However, in inflammated airways albumin from serum and DNA from lysed cells may markedly increase the viscosity of mucus. This supplement of albumin and DNA together with increased concentration and volume of glycoproteins and destruction of surface epithelium appear to be central in the impairment of mucus clearence observed in asthmatic airways.

The submucosal gland serous cells produce and release proteogly-cans as well as lactoferrin, lysozyme and secretory component to which IgA is attached (see also next paragraph). The precise role of lactoferrin (iron chelator) and lysozyme in the airway homeostasis is not known but they may process antimicrobial properties. Furthermore, these products can be used as markers of submucosal gland serous cell activation in vitro and in vivo.

Various proteases derived primarely from neutrophilic granulocytes but also from invading bacteria (including *Streptococcus pneumonia and Haemophilus influenzae*) may destroy the normal architecture of the airways. To counteract the potentially damaging effects of these pro-teases, two important protease inhibitors are present in airway mucus, namely alpha$_1$-antitrypsin and mucous protease inhibitor. Alpha$_1$-anti-trypsin is derived from plasma, whereas the mucous protease inhibitor is produced by the submucosal gland serous cells [8].

Mucus is greater than 95% (by weight) water. The hydration of mucus is regulated by sodium-potassasiun pumps of the duct cells and the surface epithelial cells [9]. In the normal airways liquid comprise a thin layer (10 μm) underneath the viscous glycoprotein-layer. The volume of this liquid layer is important for the transduction of the ciliary beat to movement of the glycoproteins. The axial movement of the mucus from the peripheral to the central airways must result in absorp-

Table 2. Constituents of mucus and their source

Source	Chemical structure
Goblet cells and submucosal gland cells	Mucous glycoproteins
Submucosal serous cells	Proteoglycans, lactoferrin, lysozyme, secretory component, mucous protease inhibitor
Capillaries in submucosa	Normal state: Water, albumin, alpha$_1$-antitrypsin Ongoing bronchial asthma: DNA, eosinophilic granulocytes, neutrophilic granulocytes, monocytes and lymphocytes

tion of a major fraction (> 99%) of the liquid. Thus airway epithelium is able to both secrete sodium to the lumen (and hence hydrate mucus) and absorb sodium (and hence dehydrate mucus). Abnormal high sodium reabsorption is seen in cystic fibrosis. Thus, dehydrated mucus is viscous and is difficult to remove from the airways. Whether mucus in asthmatic airways is dyshydrated remains to be determined.

Nervous regulation

The cholinergic nervous system innervates the submucosal glands but not the goblet cells. It is estimated that the submucosal gland cells outnumber the goblet cells by a ratio of 40 : 1 [10]. However, in the peripheral airways with a diameter of less than 1 mm there are only goblet cells. Thus, mucus production is only cholinergically innervated in the central airways.

Activation of mucus release is mediated via the muscarinic type 3 receptor. Methacholine consistently causes increased release of mucus in a variety of animal models and in humans [11, 12] (Fig. 1). Inhalation of atropine decreases the volume of mucus in the airways [13]. However, atropine may also inhibit the mucociliary clearence rate, an effect that e.g. ipratropium bromid does not prossess [14].

The adrenergic nerve-fibers also innervate the submucosal glands and both alpha- and beta-adrenergic receptors are present on these cells [15]. Alpha-adrenergic receptor activation in vitro causes a consistent increase in mucus release in all species tested, whereas beta-adrenergic receptor activation results in a more short lived and less remarkable increase in mucus release in various animal models. However, neither isoproterenol nor epinephrine enhances mucus release from human airways in vitro [12] (Fig. 1).

In humans, treatment with beta-adrenergic receptor agonists or drugs increasing the intracellular cyclic-AMP concentration (theophyllin and its derivatives) enhances the mucociliary clearence rate [16, 17]. There may be at least two explanations for this observation. The resultant bronchodilatation leads to decreased thickness of the mucus layer of the airways and hence better possibilities for a more effective transduction of the ciliary beat to the mucus. Also, increase in cyclic-AMP content of the ciliated surface epithelium cells may lead to increased frequency of the ciliary stroke.

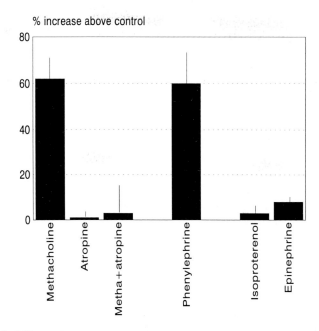

% increase above control

Fig. 1. Effects of methacholine (either alone or together with atropine), phenylephrine, isoproterenol, and epinephrine on mucus release (mean increase from control ± SEM) form human bronchi in vitro. From Shelhamer et al. [12]

There is a third nervous system in the airways entitled the non-adrenergic-non-cholinergic nervous system, since its effects can not be inhibited by antagonists of the muscarinic and the alpha- and betaadrenergic receptors [18]. The neurotransmitters of this nervous system are peptides, and thus chemically distinct from the neurotransmitters of the classic nervous systems. These so-called neuropeptides may be co-localized with the 'classic' neurotransmitters in the efferent nerves (e.g. vasoactive intestinal peptide and acetylcholine are present in the cholinergic efferent nervers), but many of the neuropeptides are situated in afferent nerves located in the surface epithelium and around the effector cells in the submocosa. Since the nerve endings of these sensory nerves may be exposed directly to irritants in the lumen of asthmatics, due to the denudation of the epithelium [4], these neuropeptides are especially interesting in the context of asthma. Activation of sensory nerve endings can lead to axon-reflexes and thus release of neuropeptides adjacent to the effector cells of the airways, including the submucosal glands [19]. Furthermore, release of these neuropeptides may lead to an inflammatory reaction [11].

The best studied neuropeptide located in the sensory nerves is substance P. Substance P stimulates the release of mucus in various species including man, substance P-containing nerves can be found around the submucosal glands, and finally receptors are demonstrated to be present on submucosal glands by autoradiography [2]. Substance P may also activate guinea pig goblet cells [20]. Since the rate of degradation of SP is lower in airways recently infected with virus, due to lower concentrations of the primary metabolizing enzyme entitled neutral endopeptidases in the submucosa, this may explain the so-called post-viral bronchial hyperreactivity syndrome [21].

An other neuropeptide presumbly located in the sensory nerves is gastrin releasing peptide (GRP). GRP can stimulate mucus release from feline airway explants [2], as well as human nasal mucosa in vivo [22]. GRP receptors are present on the submucosal glands, and GRP-containing nerves have been found around the glands and in ganglia just outsite the cartilage [2].

Several other neuropeptides have been suggested as putative neurotransmitters of the non-adrenergic-non-cholinergic nervous system including endotheline-1, endorphins and vasoactive intestinal peptide. There are stable non-peptidereceptor-antagonists under development. Until such are available for clinical use, the exact role of neuropeptides in the pathophysiology of mucus hypersecretion remains to be defined.

Eicosanoids

Airway epithelium is able to produce several of the eicosanoids including prostaglandins and leukotrienes, and these lipid-metabolites appear to be central in the regulation of the basal mucus production. The leukotrienes (especially LTC_4–LTD_4) are more potent stimulants than the prostaglandins [23–25] (Table 3).

Stimulation of the intracellular enzyme protein kinase C results in prolonged mucus hypersecretion, at least in part due to stimulation of eicosanoid production within the airway [26]. Several other mucus

Table 3. Lowest effective doses of eicosanoids causing increased mucus release from human airways in vitro[a]

Agent	Lowest dose tested resulting in increased mucus release
$PGF_{2\,alfa}$	100 µM / 200 nM
PGA_2	100 µM
15-HETE	1 nM
12-HETE	1 nM
5-HETE	1 nM
LTC_4	40 pM / 1 nM
LTD_4	40 pM / 1 nM

[a] From [23–25], Table modified from [2]. *PG* Prostaglandin; *HETE* hydroxy-eicosatetraenoic acid; *LT* leukotriene (LTC_4 and LTD_4 also named slow reacting substances of anaphylaxis)

secretagogues (see below) may mediate their stimulatory activity though the induction of the eicosanoid production in the airways.

Glucocorticoids inhibit baseline mucus production in a dose-dependent manner. Reciprocally to this inhibition, the lipocortin content of the airways increases. When lipocortin is removed by co-incubation with monoclonal antibodies, the inhibition of glucocorticoid-induced mucus production is reversed [2]. Hence, since lipocortin is known to inhibit the enzyme phospholipase A_2, responsible for the release of the eicosanoid-precursor arachidonic acid, the inhibitory effect of glucocorticoids may be mediated via inhibition of baseline eicosanoid production of the airways.

Inflammatory mediators

The mast cell, believed to be central in the development of the acute response after an allergen challange, releases a variety of mediators able to stimulate mucus release from airways, including histamine, leukotriene C_4, chymase, and platelet activating factor (PAF) [2, 27]. Stimulation of mast cell degranulation results in increased mucus release from human airways in vitro [12]. Further, Ascaris suum induced mucus hypersecretion in vitro can be inhibited by coincubation with the mast cell-degranulation-inhibitor cromolyn [28].

J. D. Lundgren

PAF induced mucus release is mediated via a specific receptor and subsequent stimulation of airway eicosanoid production [2]. However, it is controversial whether PAF plays a significant role in the pathophysiology of asthma, since trials in humans with PAF-receptor antagonist has been dissapointing so fare.

During activation of inflammatory cells oxygen free radicals are also produced, and these may stimulate mucus production via activation of prostaglandin production [29].

The most characteristic cell in asthmatic airways is the eosinophil granulocyte [4]. Eosinophils may be involved in the development of the late phase reaction. Interleukin-5 is involved in maintaining the viability of eosinophils. The predominant protein in the eosinophilic granules, the major basic protein, is cytotoxic for the surface epithelium, a destinct pathological feature of asthma pathology [4]. However, another granular protein, namely eosinophil cationic protein, has attracted even

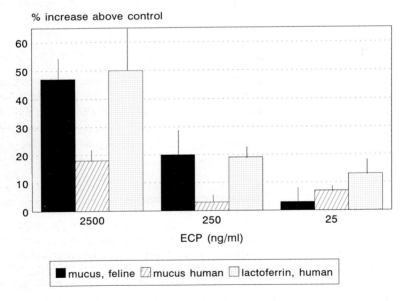

Fig. 2. Effect of eosinophil cationic protein (ECP) on mucus and lactoferrin release (mean increase from control ± SEM) form feline and human airways in vitro. From Lundgren et al. [30]

more attention since the concentration of ECP in various patient materials correlates with the clinical severity of asthma. ECP is a potent mucus secretagogue [30] (Fig. 2).

The neutrophilic granulocyte also accumulates in the airways after allergen challange [31]. Interleukin 8 and leukotriene B_4, both produced by the surface epithelium, may serve as the primary chemoattractants of this inflammatory cell. Two proteins in the neutrophil potently stimulate mucus release in vitro, namely neutrophil elastase and cathepsin G, and the effect appears to be mediated though the stimulation of eicosanoid production of the airways [32]. Furthermore, intratracheal instillation of neutrophils (or purified elastase) in rodents in vivo lead to a chronic mucus hypersecretory state due to hyperplasia of goblet cells; a phenomenon that glucocorticoid pretreatment can prevent [5]. Pathological studies of asthmatic airways demonstrates increased volume of the submucosal glands, and hyperplasia and metaplasia of goblet cells [33], and hence accumulation of neutrophils in the airways may be an important factor in causing mucus hypersecretion.

Monocytes and macrophages release several products which stimulate mucus hypersecretion [34].

Treatment strategies

To what extend pharmacological agents currently used or under development in the management of asthma affects mucus production and release from mucus producing cells of the airways is largely unknown, since there is no established method of estimating changes in the production of mucus from the lower airways in vivo. Such a method should be able to estimate production of mucus from various regions of the bronchial tree.

Four reasonable approaches to the problem of hypersecretion is presented underneath:

1. If exogenous inhalants, such as irritants, pollutants, allergens, or infectious microorganismens are present they should be removed if possible.

2. There are several receptor antagonists available for the cholinergic and adrenergic nervous system. Especially muscarinic but also alpha-adrenergic antagonists should be expected to inhibit mucus pro-

duction, whereas beta-adrenergic agonists are beneficial only by improving mucus clearence from the airways. Anti-cholinergics are less effective than betaadrenergic agonists in asthma, but the two classes of drugs may show additive effects, especially in patients with a component of irreversible bronchoobstruction [35]. A possible role of neutropeptides awaits further studies of chemical stable receptor antagonists such as the non-peptide antagonist of substance P, CP 96, 345.

3. Inhibition of inflammation appear to be the corner-stone in the management of mucus hypersecretion [2]. Anti-inflammatory therapy may inhibit mucus secretion primarily by diminishing the level of mucus secretagogues such as inflammatory chemoattractants and products released from accumulated inflammatory cells in the airway submucosa. In consequence of preventing the accumulation of eosinophils [36], damage of the airway epithelium is also avoided.

Since eicosanoids are potent mucus secretagogues, appear to be central as secondary mediators in regulating the basic and stimulated mucus production, and may also serve as chemoattractants for neutrophils, an obvious approach to manage mucus hypersecretion is to inhibit the eicosanoid production. Glucocorticoids are currently the only available effective eicosanoid inhibitor but several inhibitors of leukotriene synthesis and leukotriene receptor antagonists are currently being developed [37]. In addition to serve as an anti-inflammatory agent, glucocorticoids may also directly inhibit production of mucus [2].

Aspirin-induced asthma is thought to be induced by noncompetitive blocking of cyclooxygenase, the prostaglandins producing enzyme, by aspirin or other NSAID. Hence, increased production of eicosanoid metabolites from the other main pathway (the lipoxygenase pathway) is increased, generating large amounts of e.g. leukotrienes. Further trials with specific leukotriene inhibitors in this patient group is warrented.

The role of platelet activating factor receptor antagonists in the management of asthma awaits to be determined.

The primary mucus secretagogue from neutrophils, elastase, could be inhibited by various elastase inhibitors such as ICI 200, 355 [38].

Specific inhibition of eosinophil cationic protein and possible other of the alkaline-charged proteins of the eosinophil granules awaits to be developed. The acidic heparine was able to totally block the mucus releasing effect induced by eosinophil granules [30]. Low-dose theophylline treatment results in diminished concentrations of eosinophile cationic protein [39], and the antiinflammatory properties of theophylline [17] should be investigated further.

4. The last approach to mucus hypersecretion is improvement of its clearence. As discussed above, hydration may serve to decrease viscosity of mucus, but documentation of beneficial clinical effects is lacking. The increased viscosity of mucus in asthma is in part due to the presence of albumin and DNA. Inhalation of DNase has been demonstrated to improve the lung function and diminishes the exacerbation rate in patients with cystic fibrosis [40]. Studies in other patient groups are currently underway. Various mycolytic agents such as N-acetylcysteine is without clinical benefit in clinical trials, although certain patients appear to benefit.

In conclusion, mucus hypersecretion is an integrated part of the symptom-complex resulting in asthma, and experimental studies suggest that many of the currently used anti-asthmatic drugs probably also inhibit mucus hypersecretion. New drugs which more specifically inhibit the development of inflammation (especially the accumulation of eosinophils) are warrented.

References

1. Woolcock AJ (1994) Asthma. In: Murray JF, Nadel JA (eds) Textbook of respiratory medicine. Saunders, Philadelphia, pp 1228–1330
2. Lundgren JD, Baraniuk JN (1992) Mucus secretion and inflammation. Pulm Pharmacol 5: 81–96
3. Shimura S, Sasaki T, Sasaki H, Takishima T (1988) Bronchorrhea sputum in bronchial asthma. Am Rev Respir Dis 137: A14
4. Laitinen LA, Heino M, Laitinen A, Kava T, Haahtela T (1985) Damage of the airway epithelium and bronchial reactivity in patients with asthma. Am Rev Respir Dis 131: 138–143
5. Lundgren JD (1992) Mucus production in the lower airways (dissertation). Dan Med Bull 39: 289–303
6. James AL, Pare PD, Hogg JC (1989) The mechanics of airway narrowing in asthma. Am Rev Respir Dis 139: 242–246
7. Silberberg A (1988) Models of mucus structure. In: Braga PC, Allegra L (eds) Methods in bronchial mucology. Raven Press, New York, pp 51–62
8. Mooren HWD, Meijer CJLM, Kramps JA, Franken C, Dijkman JA (1982) Ultrastructural localization of the low molecular weight protease inhibitor in human bronchial glands. J Histochem Cytochem 30: 1130–1135

9. Boucher RC (1994) Human airway ion transport. Am J Respir Crit Care Med 150: 271–281
10. Reid LM (1960) Measurement of the bronchial mucous gland layer: a diagnostic yardstick in chronic bronchitis. Thorax 15: 132–141
11. Barnes PJ (1990) Neurogenic inflammation in airways and its modulation. Arch Int Pharmacol Ther 303: 67–82
12. Shelhamer JH, Marom Z, Kaliner M (1980) Immunologic and neuropharmacologic stimulation of mucus glycoprotein release from human airways. J Clin Invest 66: 1400–1408
13. Jeanneret-Grosjean A, King M, Michoud MC, Liote H, Amyot R (1988) Sampling technique and rheology of human trachebronchial mucus. Am Rev Respir Dis 137: 707–712
14. Pavia D, Bateman JRM, Clarke SW (1980) Deposition and clearence of inhaled particles. Clin Respir Physiol 16: 335–366
15. Barnes PJ, Basbaum CB (1983) Mapping of adrenergic receptors in the trachea by autoradiography. Exp Lung Res 5: 183–186
16. Wanner A (1988) Mucus transport in vivo. In: Braga PC, Allegra L (eds) Methods in bronchial mucology. Raven Press, New York, pp 279–290
17. Pearsson CGA (1988) Xanthines as airway anti-inflammatory drugs. J Allergy Clin Immunol 81: 615–617
18. Peatfield AC, Richardson PS (1983) Evidence for non-cholinergic, non-adrenergic nervous control of mucus secretion into the cat trachea. J Physiol 342: 335–345
19. Barnes PJ (1986) Asthma as an axon reflex. Lancet i: 242–245
20. Tokuyama K, Kuo H-P, Rohde JAL, Barnes PJ, Rogers DF (1990) Neural control of goblet cell secretion in guinea pig airways. Am J Physiol 259: L108–L115
21. Empey DW, Laitinen LA, Jacobs, L, Gold WM, Nadel JA (1976) Mechanisms of bronchial hyperreactivity in normal subjects after upper respiratory tract infection. Am Rev Respir Dis 113: 131–139
22. Baraniuk JN, Lundgren JD, Goff J, Peden D, Merida M, Shelhamer J, Kaliner M (1990) Gastrin releasing peptide (GRP) in human nasal mucosa. J Clin Invest 85: 998–1005
23. Marom Z, Shelhamer JH, Kaliner MA (1981) The effects of arachidonic acid, monohydroxyeicosatetraenoic acid, and prostaglandins on the release of mucous glycoproteins from human airways in vitro. J Clin Invest 67: 1695–1702
24. Rich B, Peatfield AC, Williams IP, Richardson PS (1984) Effects of prostaglandins E_1, E_2, and F_{2alpha} on mucin secretion from human bronchi in vitro. Thorax 39: 420–423
25. Marom Z, Shelhamer JH, Bach MK, Morton DR, Kaliner MA (1982) Slow releasing substances, leukotriene C_4 and D_4, increase the release of mucus from human airways in vitro. Am Rev Respir Dis 126: 449–451
26. Rieves RD, Lundgren JD, Logun, C, Wu T, Shelhamer J (1991) Effect of protein kinase C activating agents on respiratory glycoconjugate release from feline airways. Am J Physiol L415–L423

27. Sommerhof CP, Caughey GH, Finkbeiner WE, Lazarus SC, Basbaum CB, Nadel JA (1989) Mast cell chymase: a potent secretagogue for airway gland serous cells. J Immunol 142: 2450–2456

28. Phipps RJ, Denas SM, Wanner A (1983) Antigen stimulates glycoprotein secretion and alters ion fluxes in sheep trachea. J Appl Physiol 55: 1593–1602

29. Adler KB, Holden-Stauffer WJ, Repine JE (1990) Oxygen metabolites stimulate release of high molecular weight glycoconjugates by cell and organ cultures of rodent respiratory epithelium via an arachidonic acid-dependent mechanism. J Clin Invest 85: 75–79

30. Lundgren JD, Davey RT, Lundgren B et al (1991) Eosinophil cationic protein stimulates and major basic protein inhibits airway mucus secretion. J Allergy Clin Immunol 87: 689–698

31. Metzger WJ, Zavala D, Richardson J et al (1987) Local allergen challange and bronchial lavage of allergic asthmatic lungs: description of the model and local airway inflammation. Am Rev Respir Dis 135: 433–440

32. Lundgren JD, Rieves RD, Mullol J, Logun C, Shelhamer JH (1994) The effect of neutrophil protease enzymes on the release of mucus from feline and human airway cultures. Resp Med (in press)

33. Dunnill MS (1960) The pathology of asthma, with special reference to changes in the bronchial mucosa. J Clin Pathol 13: 27–33

34. Gollub EG, Goswami SK, Sperber K, Marom Z (1992) Isolation and characterization of a macrophage-derived high molecular weight protein involved in the regulation of mucus-like glycoconjugate secretion. J Allergy Clin Immunol 89: 696–701

35. O'Driscoll BR, Taylor RJ, Horsley MG, Chambers DU, Bernstein A (1989) Nebulized salbutamol with and without ipratropium bromide in acute airflow obstruction. Lancet i: 1418–1420

36. Laitinen LA, Laitinen A, Heino M, Haahtela T (1991) Eosinophilic airway inflammation during exacerbation of asthma and its treatment with inhaled corticosteroid. Am Rev Respir Dis 143: 423–427

37. Ford-Hutchinson AW (1991) FLAP: A novel drug target for inhibiting the synthesis of leukotrienes. Trends Pharmacol Sci 21: 68–70

38. Sommerhoff CP, Krell RD, Williams JL, Gomes BC, Strimpler AM, Nadel JA (1991) Inhibition of human neutrophil elastase by ICI 200, 355. Eur J Pharmacol 193: 153–158

39. Sullivan P, Bekir S, Jaffar Z, Page C, Jeffery P, Costello J (1994) Antiinflammatory effects of low-dose oral theophylline in atopic asthma. Lancet 343: 1006–1008

40. Fuchs HJ, Borowitz DS, Christiansen DH et al (1994) Effect of aerosolized recombinant human DNase on exacerbations of respiratory symptoms and on pulmonary function in patients with cystic fibrosis. N Engl J Med 331: 637–642

Correspondence: J. D. Lundgren, MD, DMSc, Department of Infectious Diseases (144), Hvidovre Hospital, University of Copenhagen, DK-2650 Hvidovre, Denmark.

Discussion

Moderator: If we aim on treating mucus secretion you gave us the impression that you believe only in the massive anti-inflammatory action of corticosteroids. Would you have an imagination, or a vision of other anti-inflammatory agents in the future, like theophylline that could work in the concert of anti-inflammatory actions?

Dr. Lundgren: I think that in asthma the main cause of increased mucus production is actually the influx of eosinophil granulocytes, because they are so potent in our mucous membranes to cause increased mucus production.

And as you will see later on this morning, other people will tell you that low doses of theophylline will actually prevent the influx of eosinophils into the airways and that is certainly one theoretical way that theophylline is actually preventing mucus production. But it is very important to obviously underline that you shouldn't expect these effects of theophylline within hours. It should be days or weeks, before you see these effects.

Dr. Jeffery: That was an excellent overview, very thorough. It is complex, and you've alluded to that, I think of mucus production in the form of a factory, which brings in raw materials, it stores some of its product and then it discharges it, sells it to the consumer.

And then of course you can get a multiplication of the number of factories producing the total amount of mucus, so you've got a situation where you could manipulate the raw product which goes to produce glycoprotein, you could then manipulate the discharge of that product, you could then manipulate the number of cells producing that product, you may be able to manipulate the complexing of the product with other components such as albumin. And again you alluded to that. I personally feel that it's a very important aspect for asthma. Then you've got the clearence aspect. You could manipulate clearence, which again I think is an important aspect in asthma. My question to you is, which one of those do you think is the most important to manipulate? Bearing in mind, that mucus in the correct place, in the central areas, is actually good for us. It protects the airway.

Dr. Lundgren: Well, that's a good question. I wish I had the answer, how we could pin-point which therapies we should actually be using. I think that if we can stop the cells from producing increased amounts, that would be the one thing that makes the most sense to me. You can influence the inflammatory cells and you can prevent the activatian by nerves. I think that's got to be the way to go.

Dr. Costello: Out there in the practical clinical world there's a huge volume of so-called mucolytics. And you really only refer to them in passing, like carboxymethylcystein. Do you think they should now just disappear?

Dr. Lundgren: I think definitely they should disappear. There's no evidence to support that they are working.

Dr. Costello: I think it's important in this context that that should actually be said because they probably sell more than most of the other things put together. I think in the context of a talk like this, it actually should be said that we shouldn't be prescribing them.

Moderator: Would you agree that there might be a difference between the asthmatic mucus and the mucus of the chronic bronchitic, because obviously there must be some difference in the composition of the mucus. The asthmatics have a lot of trouble to get it off their mouths, because it's so sticky and tenacious, while the mucus of the chronic bronchitic is different. And would you not believe that splitting up these disulfide bridges by acetylcystein would have some benefit?

Dr. Costello: I don't think there's any sign of evidence that they do any clinical good at all, although. I'm sure I'll find people to disagree with me.

Dr. Lundgren: I actually don't disagree. I think that there's also Dr. Jeffery who was referring to it that the viscosity of asthmatic mucus is primarily caused by albumin and DNA.

Dr. Jeffery: I think that's right. I will slightly disagree with Dr. Costello if I may, only from the experimental animal point of view. We did some work with acetylcystein, which I think was very interesting and surprising to me. In a situation of stimulating mucous cell hyperplasia and increase in the number of cells, it had a remarkable effect, whether it was given together with the irritant, which was tobacco smoke, or following cessation of tobacco smoke, it inhibited the increase in goblet cell number, and accelerated recovery to normal when cigarette smoking ceased. So I think in terms of mucolytic effect, perhaps I agree, but in terms of its effect on goblet cells and glands I think we have to be a bit cautious. It may be helpful.

Moderator: Maybe we just have to rely on the subjective benefit, if a patient says: please don't take away my mucolytic. I admit that in some cases I do prescribe mucolytic agents and the patients report grateful benefit from it.

Theophylline in the management of asthma: time for a reappraisal?

M. Weinberger

Department of Pediatrics, University of Iowa Hospital and Clinics,
Iowa City, U.S.A.

Summary

Theophylline is the oldest medication still in routine use for the treatment of asthma. Used initially as a parenteral medication for acute bronchodilatation, theophylline's primary value today results from its high degree of efficacy as a maintenance medication for preventing symptoms of chronic asthma when administered in doses that maintain serum concentrations between 10 and 20 µg/ml. Moreover, these doses that provide optimal efficacy are generally well tolerated so long as initial doses are low, increased only slowly, and serum concentration do not exceed 20 µg/ml. While newer medications have added substantially to our therapeutic armamentarium for treatment of asthma, theophylline continues to fill an important niche as the most effective oral maintenance medication and the only maintenance medication to have substantial additive effect with inhaled corticosteroids. Inhaled β-2-agonists, such as salbutamol, have improved intervention for acute symptoms but failed as reliable and safe maintenance medication. The newer longer acting agents of this class, such as salmeterol, have greater potential for use as a maintenance medication but concerns persist regarding tolerance during long term therapy. Cromolyn and nedocromil have been enthusiastically embraced by some clinicians, but a critical examination finds them weakly potent and inconvenient to use because of requirements for 4 times daily therapy. Moreover, whereas theophylline has substantial clinically important additive effects with inhaled corticosteroids, cromolyn and nedocromil do not. The empirical evidence for the clinical efficacy of theophylline as maintenance medication has recently been given additional mechanistic support by the demonstration that theophylline has anti-inflammatory in addition to bronchodilator effect. Nonetheless, it is the empirical data in the form of controlled clinical studies that provide the basis for continued use of theophylline as maintenance medication for chronic asthma. It's justifiable role is as an alternative to low dose inhaled corticosteroids as monotherapy for milder chronic asthma or as an additive maintenance medication to inhaled corticosteroids when needed.

Zusammenfassung

Theophyllin bei der Behandlung des Asthmas: Zeit für eine Neubewertung? Theophyllin ist die älteste Medikation, welche bei der Behandlung des Asthmas noch immer routinemäßig verwendet wird. Seine klinische Wirksamkeit wurde erstmals von Doktor Samson Hirsch (1922) in der deutschen Literatur demonstriert. Er behandelte zwei Asthmatiker damit und erzielte einen therapeutischen Erfolg, der den anderen zu dieser Zeit verfügbaren Medikamenten überlegen war. In den späten 30er Jahren wird in der medizinischen Literatur die intravenöse Theophyllintherapie bei akutem, epinephrin-refratärem Asthma beschrieben. Orale Theophyllinpräparate in fixer Dosiskombination mit Ephedrin wurden zur Standardbehandlung durch über 30 Jahre verwendet, nachdem die Wirksamkeit dieser Kombination in den 40er Jahren dokumentiert worden war.

Seit diesen frühen Berichten haben der massive Gebrauch von spezifischen Beta-2-Agonisten auf dem inhalativen Weg, kombiniert mit früher systemischer Kortikosteroidtherapie, den Routinegebrauch von intravenösem Theophyllin für die meisten Patienten mit akutem Asthma unnötig gemacht.

Im Gegensatz dazu bleibt die klinische Indikation für die oralen Theophylline bestehen, obwohl in stark veränderter Form seit den Berichten von Brown (1940). In einer 1974 erschienenen Publikation wurde gezeigt, daß die Kombination von Ephedrin und Theophyllin nur in puncto Toxizität synergistisch wirkte, ohne klinisch relevante additive Effekte. In dieser Arbeit wurde aber auch aufgezeigt, daß Theophyllin in einer Dosierung, welche eine Serumkonzentration von über 10 µg/ml erbrachte, besonders bei Kindern die Symptomatik erfolgreich verminderte und die Episoden von Notfallstherapie bei Kindern mit schwerem chronischen Asthma reduzierte. Die halbe Dosis, welche damals noch immer als hochdosierte Therapie angesehen wurde, hatte bei diesen Kindern einen Effekt, der nur wenig vom Placebo abwich. Darüberhinaus wurden die höheren Dosierungen gut vertragen, solang eine Serumkonzentration von 20 µg/ml nicht überschritten wurde.

In der Folge haben weitere Arbeiten die große therapeutische Sicherheit und Wirkung des Theophyllins als Erhaltungsmedikation für das chronische Asthma erwiesen, solange die Dosis durch die Messung des Serumspiegels sorgsam titriert und überwacht wird. Neuere Medikationen haben beträchtlich zu unserem therapeutischen Armamentarium in der Behandlung des Asthmas beigetragen, wie die langwirkende Beta-2-Mimetika. Diese haben zwar eine unzweifelhafte Bedeutung in der Erhaltungsmedikation erlangt, doch gibt es gewisse Vorbehalte, wenn keine begleitende inhalative Kortikosteroidtherapie stattfindet.

Cromoglykat und Nedocromil haben zwar keinen sicheren additiven Effekt, wenn sie mit inhalativen Kortikosteroiden verabreicht werden, doch ist ein solcher für Theophyllin sehr wohl nachgewiesen worden.

Die klinische Wirksamkeit des Theophyllins als Erhaltungstherapie wurde erst kürzlich durch den Nachweis der antientzündlichen Wirkung zusätzlich zum bronchodilatatorischen Effekt unterstützt. Nichtsdestoweniger sind es noch immer die Daten aus kontrollierten klinischen Studien, welche die Basis für eine Fortsetzung des Gebrauches von Theophyllin als Erhaltungstherapie beim chronischen Asthma bilden. Es erscheint gerechtfertigt, die Theophyllintherapie als Alternative zur niedrigdosierten inhalativen Kortikosteroidtherapie in Form einer Monotherapie beim milden chronischen Asthma zu verwenden, oder, wenn nötig, als eine zusätzliche Erhaltungsmedikation bei inhalativer Kortikosteroidbehandlung.

History

Theophylline is the most ancient medication still in routine use for the treatment of asthma. The first reported clinical demonstration of efficacy for theophylline appeared in the German literature in 1922 by

Dr. Samson Hirsch [1]. He treated 2 asthmatic patients and reported theophylline to be superior to other drugs in common use at the time. By the late 1930's, the medical literature indicates that intravenous theophylline was recognized to be of value for acute asthma that did not respond to epinephrine. Use of oral theophylline in a fixed dose combination with ephedrine became a standard treatment for over 30 years following an enthusiastic report of efficacy from this combination in 1940 [2]. Theophylline, whether by the parenteral route or in oral preparations, was then used predominantly to relieve acute symptoms of asthma until the early 1970's. The use of theophylline evolved following demonstration of a high degree of efficacy for theophylline as *maintenance* therapy for the prevention of asthmatic symptoms, when

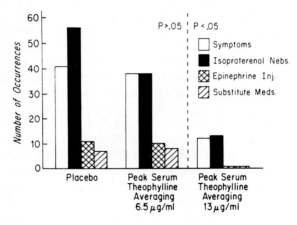

Fig. 1. Frequency and severity of asthmatic symptoms among 12 children with chronic asthma at a residential treatment center. Each patient received, in a double-blind randomized sequence, 1 week's treatment with placebo, individualized theophylline doses that resulted in peak serum concentrations averaging 13 µg/ml, and half that dose (in fixed dose combination with ephedrine as was the fashion of the time) resulting in peak serum theophylline concentrations averaging 6.5 µg/ml. Asthmatic symptoms during each 1-week period were promptly treated, when necessary, with inhaled isoprenaline (isoproterenol); if symptoms were not rapidly relieved, epinephrine was administered subcutaneously. If the patient was unresponsive to these measures, known drugs were substituted for the double-blind medications without breaking the double-blind coding (Reproduced with permission [3])

used at doses that maintained serum concentrations within a range of about 10 to 20 µg/ml (Fig. 1) [3]. Although newer drug regimens for maintenance therapy have since been introduced and extensively used, comparative studies continue to demonstrate the high degree of efficacy of theophylline for this purpose.

Comparisons with alternative antiasthmatic agents

Sodium cromoglycate (cromolyn sodium) was the first anti-asthmatic medication to be introduced *exclusively* for prevention and was marketed as the first truly prophylactic medication for asthma. However, in a multi-centered comparative trial with cromolyn, theophylline was associated with a greater frequency of asymptomatic days among children chosen because of relatively severe chronic asthma (Fig. 2) [4]. Although subsequent studies of children with *milder* disease demonstrated a similar degree of control with cromolyn or theophylline, these data were with the

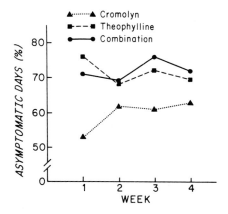

Fig. 2. Weekly frequencies of symptom-free days in 28 children with relatively severe chronic asthma at 3 centers in London and Denver. All of the children received theophylline, cromolyn, or the combination in double blind in randomized sequence. Significant drug-treatment differences were observed with the theophylline containing regimen providing significantly more asymptomatic days than cromolyn. No additive effect was apparent when the combined regimen was compared with theophylline alone (Reproduced with permission [4])

20 mg nebulizer solutions or Spinhaler formulations of cromolyn administered 4 times daily [5, 6]. Current theophylline formulations generally require only twice daily treatment, a schedule likely to be associated with much greater compliance than a 4 times daily inhaled medication. Nedocromil has recently been introduced as a cromolyn-like drug. While meaningful studies comparing nedocromil with theophylline have not been reported, the U.S.A. package insert for nedocromil contains data from a study indicating that it barely matches and certainly does not exceed the efficacy of cromolyn, even at the subpotent cromolyn dose of 2 mg from the 1 mg/puff metered dose inhaler currently marketed in the U.S. [7] (Fig. 3). Moreover, the argument that cromolyn (or nedocromil) is "anti-inflammatory" and therefore is a *priori* more appropriate than theophylline is belied by the data demonstrating that theophylline can match cromolyn's effect on the allergen induced late phase increase in airway obstruction, bronchial responsiveness, and mucosal infiltration with activated eosinophils (Fig. 4) [8, 9].

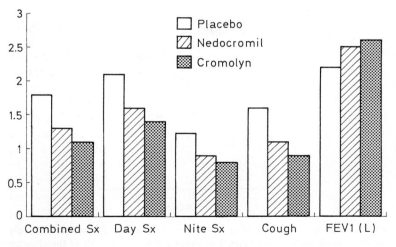

Fig. 3. Comparison of nedocromil and cromolyn among 306 patients with asthma randomly assigned placebo (n = 99), nedocromil (n = 103), or cromolyn (n = 104). Symptoms (Sx) are presented as means of daily diaries using a symptom score scale of 0 to 3; FEV$_1$ is presented as the mean in liters (Data taken from the U.S.A. package insert for Tilade brand of nedocromil)

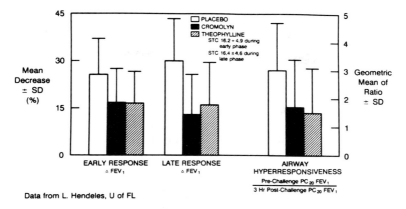

Data from L. Hendeles, U of FL

Fig. 4. Relative efficacy of theophylline and cromolyn for suppression of early and late phase response and increased airway responsiveness to histamine following allergen challenge. Theophylline matched cromolyn in this effect, which is commonly used to argue that cromolyn is anti-inflammatory [8, 57]

Comparison with oral β_2 agonists, including orciprenoline (metaproterenol) [10] and slow-release terbutaline have shown clinical advantage for theophylline, especially for nocturnal symptoms [11, 12]. Although inhaled salbutamol (albuterol) is far more potent as a bronchodilator than theophylline, theophylline nonetheless provides more stable clinical effect which is particularly important for nocturnal symptoms and other times when there will be more than 4 hours between doses of the inhaled agent [13]. The explanation for the difference in clinical effect was not an advantage of intensity of effect, where the inhaled β_2 agonist is far superior, but in sustained duration of effect. The clinical importance of this was apparent from the greater than twofold frequency of nocturnal symptoms in association with maintenance therapy during the inhaled albuterol as compared with the slow-release theophylline regimen. Data comparing theophylline with the newly released ultra-long acting β_2 agonist, salmeterol, remain limited, but tolerance to salmeterol resulting in markedly decreased ability to protect the airway from a bronchoprovocation during sustained usage is a concerning issue [14].

Theophylline has also been compared with inhaled corticosteroids. While the inhaled steroids certainly have the potential to provide a sub-

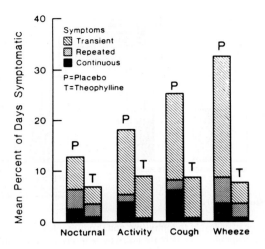

Fig. 5. Mean frequency of symptoms in 21 asthmatic children receiving a mean dose of 550 μg/day (11 puffs/day) of beclomethasone dipropionate during place-bo (P) or theophylline dosage (T) previously individualized to achieve a serum concentration of 10–20 μg/ml. The presence of nocturnal symptoms were record-ed each morning, and interference with activity, cough, and wheezing were recorded each evening as absent, transient, repeated, or continuous. Theophylline was also associated with less airway responsiveness to bronchoprovocation with treadmill exercise and fewer interventions with both inhaled β₂ agonists and daily oral corticosteroids (Reprinted with permission [16])

stantial degree of clinical efficacy, theophylline compared favorably with a study utilizing lower doses of inhaled corticosteroids in "mild to mod-erate" asthmatics [15]. Moreover, theophylline has substantial additive effect with both an inhaled corticosteroid (Fig. 5) and an alternate morn-ing oral prednisone maintenance regimen [16, 17]. In contrast, all 3 pub-lished clinical trials of cromolyn examining the potential for additive effect with inhaled corticosteroids were impressively negative [18–20].

Considerations for theophylline use in clinical practice

Pharmacodynamics – relationship of dose to efficacy & toxicity: The degree of clinical effect from theophylline described in the above studies is most readily seen when serum concentrations are maintained between 10 and 20 µg/ml. Response of the airways can be demonstrated to parallel changes in serum concentration (Fig. 6) [21, 22]. Similar to the relationship of serum concentration to bronchodilator activity, effect on stabilizing the responsiveness of the asthmatic airways to exercise can also be demonstrated to relate closely to serum concentration [23]. This relationship of bronchodilatation and airway responsiveness to serum concentration is reflected in the relationship between serum concentration and symptoms. Lower serum concentrations were associated with little clinical effect compared with placebo in our earlier study, while serum concentrations over 10 µg/ml largely "turned off" the asthma (Fig. 1). A more recent study has confirmed relationship between suppression of symptoms and serum concentrations reported in earlier studies (Fig. 7).

On the other hand, theophylline clearly has the greatest potential for serious toxicity of any medication used for asthma. An extensive review

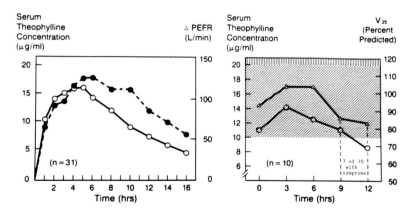

Fig. 6. Relationship between serum theophylline concentration and pulmonary function. Data from the left graph were obtained from 31 adults who received a single 7.5 mg/kg dose of a rapid-release plain tablet. Pulmonary function closely paralleled the rise and fall in serum concentration [21]. Data from the right graph were obtained from 10 children receiving a slow-release theophylline preparation every 12 hours. Symptoms of asthma requiring inhaled albuterol occurred in 7 of the 10 children exclusively during the last 3 hours of the dosing interval when serum concentrations were generally falling below 10 µg/ml; this intervention blunted further fall in pulmonary function [22] (Reproduced with permission)

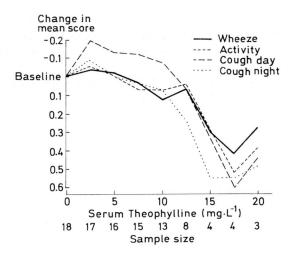

Fig. 7. Relationship between 4 hour post-dose peak serum theophylline concentrations during slow-release theophylline administration and change in mean symptom scores taken from daily diaries among 20 children treated with three different levels of maintenance theophylline for 3 weeks each in a cross-over manner (Reproduced with permission [62])

of the world's English language literature of reported cases of theophylline toxicity is recorded elsewhere [24]. However, toxicity is not idiosyncratic. Minor side effects of theophylline, common when dosage is initiated too aggressively, are avoided by slow clinical titration that begins with low doses and increases dosage only slowly as tolerated with final dose guided by serum concentration [25]. This permits tolerance to develop for the caffeine-like side effects that are commonly present during more agressive initial therapy. Serious toxicity relates to serum concentration and increases in likelihood and degree as serum concentration exceeds 20 mcg/ml.

Concern has sometimes been raised regarding the long-term effects of maintenance therapy with a systemic medication such as theophylline, particularly in growing children. Both neuropsychological and metabolic effects of theophylline during maintenance therapy appear to be similar to those associated with caffeine [26]. They include improved performance on tasks requiring mental alertness such as remembering a number sequences, slightly decreased motor steadiness, and a tendency towards increased frequency of various affective and somatic responses on a questionnaire designed to detect mood and behavior changes. While statistically significant, the differences observed were quantitatively small, and none of this was detectable on daily symptom diaries, even over extended periods, suggesting that these findings are generally of limited clinical importance for most patients.

Concerns regarding the potential for effects of theophylline on school performance and behavior alarmed the public and some physicians, but a critical examination of this controversy found inadequate data to support the concerns [27]. Subsequently, definitive studies on both cognitive performance and behavior found no evidence for effects of theophylline on either performance on standardized academic achievement tests or behavioral even when parents reported their impressions of such associations [28, 29].

Pharmacokinetics – Influence of absorption, distribution, and elimination of theophylline on clinical use: Theophylline is rapidly, consistently, and completely absorbed from oral liquids and plain uncoated tablets [30]. It then distributes throughout the extracellular space and to a lesser extent to the intracellular space with an apparent volume of distribution averaging about 0.5 l/kg in both children and adults [31]. Protein binding is only about 40% [32]. There is considerable interpatient variation in the rate of theophylline metabolism and consequent elimination. This results in at least a 4 fold range in weight adjusted dosage requirements [33, 34]. Despite the large *inter*patient variability of theophylline elimination, *intra*patient variability in elimination rate appears to be relatively small in the absence of confounding variables [35].

Formulation – Influence of product on absorption: While theophylline is still used occasionally for acute therapy where the rapid onset of action of the intravenous route may be indicated, oral treatment has become largely targeted at maintenance prophylactic therapy where the sustained action of slow-release theophylline preparations has been desirable. During the 50 years of clinical usage of theophylline, many formulations have been marketed as companies vie for perceived commercial advantage. Many of these preparations were chosen because of marketing perceptions rather than scientific justification. Some formulation selections may also have been a result of misguided notions of the chemistry of theophylline. Certainly, the past use of the many so-called "salts" of theophylline had little rationale. Even the traditional use of

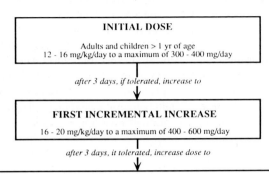

INITIAL DOSE

Adults and children > 1 yr of age
12 - 16 mg/kg/day to a maximum of 300 - 400 mg/day

after 3 days, if tolerated, increase to

FIRST INCREMENTAL INCREASE

16 - 20 mg/kg/day to a maximum of 400 - 600 mg/day

after 3 days, it tolerated, increase dose to

HIGHEST DOSAGE BEFORE MEASUREMENT OF SERUM CONCENTRATION

18 - 22 mg/kg/day to a maximum of 600 - 800 mg/day

Check serum concentration at the time of estimated peak serum concentration for the formulation used when no doses have been missed, added, or taken at unequal intervals for 3 days.

SERUM CONCENTRATION	DIRECTIONS
Below 7.5 µg/ml	Increase dose about 25% and recheck serum theophylline for guidance in further dose adjustment.
7.5 to 10 µg/ml	Increase dose about 25% and maintain *if tolerated*.
10 to 20 µg/ml	**Maintain dose *if tolerated*.***
20 to 25 µg/ml	Decrease dose at least 10%.
25 to 30 µg/ml	Skip next dose and decrease subsequent doses at least 25%.
Over 30 µg/ml	Skip next 2 doses, decrease subsequent doses at least 50%, and recheck serum theophylline for guidance in further dose adjustment.

ONCE FINAL DOSAGE IS DETERMINED BASED ON NEED, TOLERANCE, AND SERUM CONCENTRATION, RECHECK AT 6 TO 12 MONTH INTERVALS (DEPENDING ON RATE OF GROWTH) UNLESS CLINICALLY INDICATED SOONER

*Finer adjustments in dosage within this range of serum concentrations may be needed for some patients; drug interactions or physiologic abnormalities may mandate earlier reexamination.

"aminophylline," the so-called ethylenediamine salt of theophylline, for parenteral use has no scientific justification. Theophylline parenteral preparations without ethylenediamine have been readily available in the U.S. for over 10 years. These avoid the occasional allergic reactions that have been associated with the ethylenediamine component [36–41] and have become the parenteral preparations of choice.

The slow-release formulations are today the most clinically relevant preparations for maintenance therapy, even for young children where bead-filled capsules have been extensively used opened and "sprinkled" on spoonfuls of soft food for administration to young children. Because of their widespread clinical use, many unique formulations have been developed and marketed under an even wider variety of brand names. Despite commercial claims for formulation-specific dosing intervals, it is the characteristic rate of absorption of the product combined with the rate of elimination of the patient at a selected dosing interval that results in the consequent fluctuations in serum concentration. The methodology for evaluating products is extensively reviewed elsewhere [42, 43]. While absorption of theophylline from all products can be influenced by circadian variations in gastric emptying [44], food has specific effects on some formulations that may result in adverse consequences. Formulations with large variability in extent of absorption or where rate of absorption is affected to a major degree should not be used because of increased risk of toxicity or excessive variations in efficacy [45–55].

One of the more recent developments in slow-release theophylline marketing has been the commercial obfuscation associated with claims for once daily dosing [56, 57]. Regardless of marketing claims, how-

←————————————————————————————————

Fig. 8. Guidelines for slow clinical titration of theophylline dosage with final dosage adjustment based on the results of a serum concentration measurement. For infants < 1 year of age, the initial daily dosage can be calculated by the following regression equation: Dose (in milligrams per kilogram per day) = (0.2) (age in weeks) + 5.0; subsequent dosage increases in this age group should be based on a peak serum concentration measurement at least 3 days after the start of therapy. Whenever side effects occur, dosage should be reduced to a previously tolerated lower dose. In the presence of risk factors for toxic effects (e.g., fever, erythromycin therapy, or oral contraceptive therapy), the "initial dose" should not be exceeded without first documenting that a reliably measured steady-state peak concentration is < 10 µg/ml. In patients without risk factors, the subsequent dosage increases are made if judged to be clinically indicated [8, 57, 63].

ever, the principles of product selection remain the same. The rate and extent of absorption need to be identified, and the predicted fluctuations in serum concentration for patients with defined rates of elimination can then be predicted as previously described [42].

A practical approach to safe and effective dosing of theophylline: Selection of dosage for theophylline must consider rate of theophylline elimination which varies, on average, with age, and the large variability even within age groups is mirrored by the consequent wide range in dosage. An additional confounding factor that influences dosing schedule is the relationship between complaints from theophylline use and the rate at which the therapeutic range is approached. The frequency of minor adverse effects is greatly diminished when initial dosage is low and the final therapeutic dose is approached slowly by clinical titration over a 1- to 2-week period, with final dose guided by measurement of serum concentration (Fig. 8). When this dosage scheme is used, the frequency of even minor adverse effects detectable by history is only about 2% in children and only somewhat higher in adults, so long as serum concentrations are under 20 µg/ml [25].

Once final dosage is established based on guidance by measurement of serum concentration, dosage requirements generally remain stable for extended periods. However, sustained fever, heart failure, liver disease, some macrolide antibiotics, some quinoline antibiotics, and cimetidine slow theophylline elimination and thereby cause increases in steady-state serum concentration. On the other hand, phenytoin, phenobarbital, and cigarette (or marijuana) smoking will increase theophylline elimination, thereby decreasing steady-state serum concentrations. (It is important to consider that patients initially titrated to appropriate dosage based on serum concentration while smoking or receiving phenytoin or phenobarbital will have slower theophylline elimination and increased serum concentrations should these exogenous factors be discontinued.) Drug interactions with theophylline of potential clinical relevance are summarized elsewhere [58], but it is essential to consider that any drug influencing the hepatic P_{450} enzyme system has the potential to influence theophylline elimination and thereby alter dose requirements.

Despite the potential for toxicity and the concerning frequency of dosing errors in some institutions [59], an investigation of extensive clinical use in a community setting found an exceedingly low frequency of serious toxicity [60]. Among 225,000 theophylline prescriptions for 36,000 patients, only 2 patients (1 child and 1 adult) developed theophylline-induced seizures; it was not indicated whether these were from prescribed doses or ingestions. Thus, iatrogenic theophylline toxicity can be avoided. The following precautions can ensure optimal safe use of theophylline:

- Always begin with a low dose and titrate slowly; doses should be maintained only if tolerated.
- Always guide final dosage by measurement of serum theophylline concentration during the time of estimated peak serum concentration.
- Instruct the patient to always hold a dose of theophylline until a serum concentration can be measured if there is any suggestion of adverse effects, especially persistent headache, nervousness, irritability, tachycardia, nausea, or vomiting.
- Reduce dosage by 1/2 for fever (temp > 38.5°C) sustained beyond 24 hours. Reduce maintenance dose by 1/3 whenever erythromycin or ciprofloxacin is added and by 1/2 for cimetidine, troleandomycin, and oral contraceptives (a smaller decrease may be adequate for very low estrogen-containing products); recheck theophylline level and readjust the dosage if these are to be maintained; reduce dose by 1/2 if maintenance therapy with an enzyme inducer such as carbamazepine or phenytoin is discontinued; monitor theophylline levels monthly if cigarette smoking is stopped; consider the possibility of other medications and physiological abnormalities affecting theophylline elimination and thereby altering maintenance dose requirements.
- Provide adequate instruction (preferably as printed instruction forms) for patient and/or family to understand benefits and risks of theophylline [61].

Where does theophylline fit in a rational, effective, and efficient scheme for managing asthma?

We have several medications demonstrated to be efficacious for asthma. But *efficacy* alone is not sufficient. The medications must also be *effective* in clinical use. Furthermore, efficiency of treatment must be considered. Is the treatment cost-effective, not just in terms of actual monetary cost but in patient time and effort which influences compliance, which in turn influence effectiveness and outcome? In considering medications for asthma, we must also consider the therapeutic strategy desired. We use medications for *intervention,* i.e. to stop acute symptoms of asthma when they occur, and we use medications for *maintenance,* i.e. to prevent symptoms of asthma for patients with sufficiently frequent symptoms to justify continuous medication. Theophylline's traditional use as intervention has largely been supplanted by the current generation of inhaled β_2 agonists which can safely be pushed to considerably higher doses than older adrenergic bronchodilators. However, consideration of the various alternative maintenance regimens argues for inhaled corticosteroids and theophylline being the primary agents. Oral β_2 agonists are not as effective. Inhaled β_2 agonists available prior to salmeterol had insufficient duration of action to be valuable as maintenance.

Salmeterol shows promise but comparative data remains sparse and the issue of tolerance is not resolved. Cromolyn, while well documented in 20 mg doses by Spinhaler or nebulizer solution administered 4 times daily to be effective in controlling patients with milder degrees of chronic asthma, is the most expensive and least convenient of the alternatives. Four times daily administration is simply not realistic long-term medication for a child in school, at least when the twice

Table 1. An efficient method for sequential selection of pharmacologic agents based on the pharmacologic potential of the medication, the physiologic abnormality, and the therapeutic strategy

Physiologic abnormality	Treatment strategy	
	Intervention	Maintenance (if indicated)
Bronchodilator response complete and sustained	Inhaled beta₂ agonist	Theophylline *or* inhaled corticosteroid
Bronchodilator subresponsive	Short course of high dose oral corticosteroid	Theophylline *and* inhaled corticosteroids *or* theophylline and alternate-morning prednisone

daily alternative of low dose inhaled corticosteroids or theophylline offer maintenance medication with a higher efficacy yield. When more than one medication is needed, the only combinations shown to have additive effect in controlled studies are inhaled corticosteroids and theophylline (or oral alternate morning corticosteroids and theophylline) [16, 17]. This approach to therapeutic decision making is summarized in Table 1.

References

1. Hirsch S (1922) Klinischer und experimenteller Beitrag zur krampflösenden Wirkung der Purinderivate. Klin Wochenschr 1: 615–618
2. Brown EA (1940) New type of medication (ephedrine, phenobarbital and theophyllin) to be used in bronchial asthma. N Engl J Med 223: 843–846
3. Weinberger M (1978) Theophylline for treatment of asthma. J Pediatr 92: 1–7
4. Hambleton G, Weinberger M, Taylor J, Cavanaugh M, Ginchansky E, Godfrey S, Tooly M, Bell T, Greenberg S (1977) Comparison of cromoglycate (cromolyn) and theophylline in controlling symptoms of chronic asthma. Lancet 1 : 381–385
5. Shapiro GG, Konig P (1985) Cromolyn sodium: a review. Pharmacotherapy 5: 156–170
6. Weinberger M (1985) Commentary – Cromolyn sodium: A review. Pharmacotherapy 5: 169–170
7. Patel KR, Wall RT (1986) Dose-duration effect of sodium cromoglycate aerosol in exercise-induced asthma. Eur J Respir Dis 69: 256–260
8. Hendeles L, Harman E, Huang D, Cooper R, Delafuente J, Blake K (1991) Theophylline attenuation of allergen-induced airway hyper-reactivity and late response. J Allergy Clin Immunol (in press) (abstract published J Allergy Clin Immunol) 87: 167
9. Sullivan P, Songul B, Zeina J, Page C, Jeffery P, Costello J. Anti-inflammatory effects of low-dose oral theophylline in atopic asthma. Lancet 343: 1006–1008
10. Dusdieker L, Green M, Smith GD, Ekwo EE, Weinberger M (1981) Comparison of orally administered metaproterenol and theophylline in the control of chronic asthma. J Pediatr 101: 281–287
11. Wilkens JH, Wilkens H, Heins M, Kurtin L, Oellerich M, Sybrecht GW (1987) Treatment of nocturnal asthma: the role of sustained-release theophylline and oral beta-2-mimetics. Chronobiol Int 4: 387–396
12. Heins M, Kurtin L, Oellerich M, Maes R, Sybrecht GW. Nocturnal asthma: slow-release terbutaline versus slow-release theophylline therapy. Eur Respir J 1: 306–310
13. Joad J, Ahrens RC, Lindgren SD, Weinberger MM (1986) Relative efficacy of maintenance therapy with theophylline, inhaled albuterol, and the combination for chronic asthma. J Allergy Clin Immunol 79: 78–85

14. Cheung D, Timmers MC, Zwinderman AH, Bel EH, Dijkman JH, Sterk PJ (1992) Long term effects of a long acting β_2-adrenal receptor agonist, salmeterol, on airway hyperresponsiveness in patients with mild asthma. N Engl J Med 327: 1198–1203

15. Tinkelman DG, Reed CE, Nelson HS, Offord KP (1993) Aerosol beclomethasone dipropionate compared with theophylline as primary treatment of chronic, mild to moderately severe asthma in children. Pediatrics 92: 64–77

16. Nassif EG, Weinberger MM, Thompson R, Huntley W (1981) The value of maintenance theophylline for steroid dependent asthma. N Engl J Med 304: 71–75

17. Brenner M Berkowitz R Marshall N Strunk RC (1988) Need for theophylline in severe steroid-requiring asthmatics. Clin Allergy 18: 143–150

18. Toogood JH, Jennings B, Lefcoe NM (1981) A clinical trial of combined cromolyn/beclomethasone treatmen for chronic asthma. J Allergy Clin Immunol 67: 317–324

19. Hiller EF, Milner AD (1975) Betamethasone 17 valerate aerosol and disodium cromoglycate in severe childhood asthma. Br J Dis Chest 69: 103–106

20. Dawood AG, Hendry AT, Walker SR (1977) The combined use of betamethasone valerate and sodium cromoglycate in the treatment of asthma. Clin Allergy 7: 161–165

21. Richer C, Mathier M, Bah H, Thuillez C, Duroux P, Giudicelli JF (1982) Theophylline kinetics and ventilatory flow in bronchial asthma and chronic airflow obstruction. Clin Pharmacol Ther 31: 579–586

22. Simons FEF, Lucriuk GH, Simons KJ (1982) Sustained-release theophylline for treatment of asthma in preschool children. Am J Dis Child 136: 790–793

23. Pollock J, Kiechel F, Cooper D, Weinberger M (1977) Relationship of serum theophylline concentration to inhibition of exercise-induced bronchospasm and comparison with cromolyn. Pediatrics 60: 840–844

24. Hendeles L, Weinberger M (1983) Theophylline. In: Middleton E, Reed C, Ellis E (eds) Allergy: principles and practice, 2nd edn. CV Mosby, St. Louis, Missouri, pp 535–574

25. Milavetz G, Vaughan L, Weinberger M, Hendeles L (1986) Evaluation of a scheme for establishing and maintaining dosage of theophylline in ambulatory patients with chronic asthma. J Pediatr 109: 351–354

26. Joad J, Ahrens R, Lindgren S, Weinberger M (1986) Extrapulmonary effects of maintenance therapy with theophylline and inhaled albuterol in patients with chronic asthma. J Allergy Clin Immunol 78: 1147–1153

27. Weinberger M, Lindgren S, Bender B, Lerner J, Szefler S (1987) Effects of theophylline on learning and behavior: reason for concern or concern with reason? J Pediatr 111: 471–474

28. Bender B, Milgrom H (1992) Theophylline-induced behavior change in children: an objective evaluation of parents' perception. JAMA 267: 2621–2624

29. Lindgren S, Lokshin B, Stromquist A, Weinberger M, Nassif E, McCubbin M, Frasher R (1992) Does asthma or its treatment with theophylline limit academic performance in children? N Engl Med 327: 926–930

30. Hendeles L, Weinberger M, Bighley L (1977) Absolute bioavailability of oral theophylline. Am J Hosp Pharm 34: 525–527

31. Hendeles L, Weinberger M (1982) Theophylline: A "state-of-the-art" review. Pharmacotherapy 3: 244

32. Shaw LM, Fields L, Mayock R (1982) Factors influencing theophylline serum protein binding. Clin Pharmacol Ther 32: 490–496

33. Ginchansky E, Weinberger M (1977) Relationship of theophylline clearance to oral dosage in children with chronic asthma. J Pediatr 91: 655–660

34. Wyatt R, Weinberger M, Hendeles L (1978) Oral theophylline dosage for the management of chronic asthma. J Pediatr 92: 125–130

35. Milavetz G, Vaughan L, Weinberger M. Stability of theophylline elimination rate. Clin Pharmacol Ther 41: 388–391

36. Petrozzi JW, Shore RN (1976) Generalized exfoliative dermatitis from ethylenediamine. Arch Dermatol 112: 525–526

37. Folli HL, Cupit GC (1978) Ethylenediamine hypersensitivity. Drug Intell Clin Pharm 12: 482–483

38. Elias JA, Levinson AI (1981) Hypersensitivity reactions to ethylenediamine in aminophylline. Am Rev Respir Dis 123: 550–552

39. Provost TT, Jillson OF (1967) Ethylenediamine contact dermatitis. Arch Dermatol 96: 231–234

40. de la Hoz B, Perez C, Tejedor MA, Lazaro M, Salazar F, Cuevas M (1993) Immediate adverse reaction to aminophylline. Ann Allergy 71: 452–454

41. Weinberger M (1993) Why adulterate theophylline? Guest editorial. Ann Allergy 71: 419

42. Hendeles L, Iafrate RP, Weinberger M (1984) A clinical and pharmacokinetic basis for the selection and use of slow-release theophylline products. Clin Pharmacokinet 9: 95–135

43. Weinberger M, Hendeles L (1983) Slow-release theophylline – rationale and basis for product selection. N Engl J Med 308: 760–764

44. Goo RH, Moore JG, Greenberg E, Alazraki NP (1987) Circadian variation in gastric emptying of meols in humans. Gastroenterology 93: 515–518

45. Hendeles L, Weinberger M, Milavetz G, Hill M, Vaughan L (1985) Food-induced dose dumping from a once-a-day'theophylline product as a cause of theophylline toxicity. Chest 87: 758–765

46. Pedersen S, Moller-Petersen J (1984) Erratic absorption of a slow-release theophylline sprinkle product caused by food. Pediatrics 74: 534–538

47. Weinberger M, Milavetz G (1986) Influence of formulation on oral drug delivery: Considerations for generic substitution and selection of slow-release products. Iowa Med 76: 24–28

48. Sips AP, Adelbroek PM, Kulstad S, deWolff FA, Dijkman JH. Food does not affect bioavailability of theophylline from Theolin Retard. Eur J Clin Pharmacol 1984; 26: 405–407

49. Welling PG, Lyons LL, Craig WA, Trochta GA (1975) Influence of diet and fluid of bioavailability of theophylline. Clin Pharmacol Ther 17: 475–480

50. Pedersen S, Moller-Petersen J (1982) Influence of food on the absorption rate

and bioavailability of a sustained release theophylline preparation. Allergy 37: 531–534

51. Osman MA, Patel RB, Irwin DS, Welling PG (1983) Absorption of theophylline from enteric coated and sustained release formulations in fasted and nonfasted subjects. Biopharm Drug Dispos 4: 63–72

52. Leeds NH, Gal P, Purohit AA, Walter JB (1982) Effect of food on the bioavailability and pattern of release of a sustained-release theophylline tablet. J Clin Pharmacol 22: 196–200

53. Lagas M, Jonkman JHG (1983) Greatly enhanced bioavailability of theophylline on postprandial administration of a sustained release tablet. Eur J Clin Pharmacol 24: 761–767

54. Karim A, Burns T, Wearley L, Streicher J, Palmer M (1985) Food-induced changes in theophylline absorption from controlled-release formulations. Part I. Substantial increased and decreased absorption with Uniphyl tablets and Theo-Dur Sprinkle. Clin Pharmacol Ther 38: 77–83

55. Milavetz G, Vaughan L, Weinberger M, Hendeles L (1987) Bioavailability of oral theophylline: single and multiple dose studies of Uniphyl. J Allergy Clin Immunol 80: 723–729

56. Weinberger M. Theophylline QID, TID, BID, and now QD? (1984) A report on 24-hour dosing with slowrelease theophylline formulations with emphasis on analysis of data used to obtain food and drug administration approval for Theo-24. Pharmacotherapy 4: 181–198

57. Weinberger M (1986) Clinical and pharmacokinetic concepts of 24-hour dosing with theophylline. Ann Allergy 56: 2–8

58. Weinberger M, Hendeles L (1993) Theophylline. In: Middleton E, Ellis E (eds) Allergy: principles and practice, 4th edn. C. V. Mosby, St. Louis, pp 816–855

59. Schiff GD, Hegde HK, LaCloche L, Hryhorczuk DO (1991) Inpatient theophylline toxicity: preventable factors. Ann Intern Med 114: 748–753

60. Derby LE, Jick SS, Langlois JC, Johnson LE, Jick H (1990) Hospital admission for xanthine toxicity. Pharmacotherapy 10: 112–113

61. Weinberger M (1990) Managing asthma, Appendix B – Patient educational material. Williams & Wilkins, Baltimore, pp 275–282

62. Neijens JH, Duiverrnan EJ, Graatsma BH, Kerrebijn KF (1985) Clinical and bronchodilating efficacy of controlled-release theophylline as a function of itsi serum concentrations in preschool children. J Pediatr 107: 811–815

63. Hendeles L, Weinberger M, Szefler S, Ellis E (1992) Safety and efficacy of theophylline in children with asthma. J Pediatr 120: 177–183

Correspondence: M. Weinberger, M.D., Department of Pediatrics, University of Iowa Hospital and Clinics, 200 Hawkins Drive, Iowa City, IA 52242, U.S.A.

Discussion

Moderator: Thank you Dr. Weinberger. And as expected, the problems, the ideas, visions and experiences of Dr. Weinberger would fill a symposium that would last for 3 or 4 days. And I congratulate you for this very precise and concise presentation. Dr. Weinberger's paper is open for discussion.

N.N.: You mentioned the Tinkelman study and showed that the control of asthma was comparable between theophylline on the one side and steroids on the other side. But you said they didn't dose theophylline in the optimal way. What could they have done better?

Dr. Weinberger: They had a peculiar aspect of the protocol with regard to the use of theophylline in that they would first determine the appropriate dose that attained therapeutic serum concentrations and then permitted the different centres to lower the dose if they chose without any sound rationale.

So the serum concentrations were not on average in the 10 to 20 microgram per ml range, it was somewhat lower, closer to what we had shown to be not much different from placebo. Some patients were in the therapeutic range of 10 to 20 µg/ml and some weren't. There was a lack of consistency among the multiple centers involved in this study that was allowed by the protocol.

N.N.: I agree with that. First of all the point should be emphasized that theophylline came out as equally effective as inhaled steroids in this group of children. And secondly there were fewer drop-outs from the theophylline arm of the study despite the fact, as Miles Weinberger says, it wasn't conducted ideally.

Dr. Poulter: Given all the wonderful information of the 20 years you've just described, how do you explain the reduction in the use of theophylline particularly in the U.K., over the last 10 years or so? Was there another data that perhaps doesn't quite confirm all the things that you've actually described?

The second question would be that with the more recent information that there may be some anti-inflammatory effect of theophyllines, I would guess it's unlikely to simply go back and use them in the U.K. as they had been used in the past, for a variety of reasons. Could you suggest other ways in which they might be applied therapeutically? In the light of the new information regarding their activity as anti-inflammatories?

Dr. Weinberger: You're asking me to dive into the psychology of physician decision-making, which is very difficult. A great deal of medical practice is based on medical custom, which has a very strong influence. I remember when I first presented our data related to theophylline to the ... Academy .... in the United States, in 1972, I met with a great deal of resistance because everyone was defending the ephedrin theophylline combination products, saying this was great stuff, and why should we change? Well of course most people did change, the ephedrine-theophylline combinations were still around in a residual manner, but are no longer common practice. Part of the different evolvement in the U.K. and the U.S. was because just about the same time we published our data on the prophylactic value of theophylline, cromolyn had become available as the first prophylactic agent for the treatment of asthma in the U.K., 5 years ahead of where we had it in the U.S. And there was a lot of enthusiasm for cromolyn when it first came out in the U.K. However, maybe there was resistance to the use of theophylline, although Dr. Simon Godfrey, pediatric pulmonologist in the U.K. at the time was involved with a lot of the studies of cromolyn, and was a strong proponent of its use. When I visited him a few years later at the Hadass Medical Center in Jerusalem, his fellows acknowledged, although Simon wouldn't, that they were using more theophylline than cromolyn. I have difficulty understanding what drives things in the U.K. since I'm not part of that the medical social scene.

In the U.S. there's been a decreased use of theophylline in recent years because of publicity scares regarding its use. Largely commercially driven by business as a way of marketing cromolyn. So these are multifactorial things, what actually leads to position usage.

Moderator: Could you picture that the easier availability of inhalative corticosteroids in Great Britain in terms of costs, and the higher dosage, could have contributed to the reduction of the prescription of theophylline in the U.K.?

Dr. Weinberger: In the U.S., before cromolyn ever became available, there was rapid advancement in the development of theophylline products. These were much slower to get on the U.K. market. In fact, the distribution rights of Theodur, the market leader in the U.S. for many years, were purchased in the U.K. by Fisons, and then they never marketed it. That might have had an influence on the U.K.

Moderator: This situation impresses me like the story of the two girls with two dolls. The rich girl has a fancy doll, furnished with computers, talking and peeing, and doing everything that you can picture, and this doll is called inhalative corticosteroids. And there is this poor girl who owns a doll just made from a piece of wood, which is called theophylline. But the girl with the piece of wood paints a face on the wood, fixes some colourful stuff around the piece of wood, takes it in her arms and sings a lullaby, and sooner or later the girl with the fancy doll will look over and finally come over and play with the piece of wood and get all excited about it. But, that's the crucial point, all of a sudden the poor girl sneaks over and starts playing with the lonesome fantastic doll.

I think it's good that children can play with both dolls at the same time. And this is one of the purposes of this symposium. And as you beautifully lined out, the combination of the drugs is very potent, very practicable and reduces probably side-effects of both.

The effects of theophylline on airway inflammation in asthma

J. Costello

King's College, School of Medicine & Dentistry, Department of Thoracic Medicine, London, U.K.

Summary

Although the precise mode of anti-inflammatory action of theophylline is still not certain, it is likely that the effects are being achieved through phosphodiesterase inhibition. We should now apply our new knowledge of these actions of theophylline to maximise the anti-inflammatory benefit for asthmatic subjects. This might include placing children on long-term low dose theophylline, and carries the particular advantage of being an oral medication ensuring better compliance. It is possible that this form of therapy could prevent the structural change in the airway which is seen in untreated chronic asthmatics. If these data are substantiated a review of the position of theophylline in asthma guidelines might be indicated.

Zusammenfassung

Die antientzündliche Wirkung des Theophyllins. Theophyllin ist ein vielerorts verschriebenes Medikament gegen Asthma, üblicherweise in der Form von oralen Retardverbindungen. Obwohl seine Wirkung schon vor über 100 Jahren erkannt wurde, ist der Modus seiner Wirkung noch immer nicht klar, weder auf zellulärem noch auf klinischem Niveau. Die vermuteten zellulären Mechanismen seiner Wirkung inkludieren die Hemmung des zyklischen Nukleotids Phosphodiesterase,

den Antagonismus gegen Adenosin und gegen verschiedene andere
Katecholamineffekte, den Phospholipidmetabolismus und die Calzium-
verfügbarkeit und Utilisation innerhalb der Zelle.

Klinisch relaxiert Theophyllin die Bronchialmuskulatur, man weiß
aber auch, daß es den mukoziliären Transport verbessert, sowie auch die
Effizienz der diaphragmalen Kontraktion. Ferner hat es eine Wirkung
als zentralangreifendes Stimulans für die Atmung.

Sicher ist, daß der bronchodilatatorische Effekt alleine nicht seine
lange Wirkung erklären würde. Andererseits scheint eine antiinflammatorische Wirkung des Theophyllins aus zahlreichen jüngeren Arbeiten
hervorzugehen.

Wir überblicken bereits eine große Zahl von invitro, tierexperimen-
tellen und humanen Daten, welche die antiinflammatorische und im-
munmodulatorische Wirkung der Theophylline untermauern. Die In-
vitrostudien betreffen Lymphozyten und insbesondere T-Lymphozyten
aus dem peripheren Blut. Es sind Effekte auf die Bildung von Leukotrie-
nen (B4) aus kultivierten humanen Monozyten und auf die Aktivierung
von polymorphkernigen Zellen beschrieben worden. Im Tiermodell
vermindert das Theophyllin die Atemwegsentzündung und Hyperreaktivität im Rattenmodell und schwächt die späte asthmatische Reaktion
(LAR) beim Meerschweinchen ab. Theophyllin ist ein potenter Hem-
mer der Hyperreaktivität auf Ovalbuminsensibilisierung (neugeborene
Kaninchen).

Beim Menschen kennen wir nunmehr verschiedene Arbeiten, welche überzeugend die Hemmwirkung von Theophyllin auf die LAR
beschreiben, sowohl als einzelne intravenöse Dosis, als auch oral. Wenn
man die LAR als Modell der Atemwegsentzündung beim Asthma be-
trachtet, dann würde dies einer antiinflammatorischen Wirkung des
Theophyllins entsprechen.

Zwei Studien, welche wir in unseren Laboratorien durchgeführt
haben, haben einen überzeugenden Effekt des Theophyllins nicht nur
auf die LAR, sondern auch auf entzündliche Infiltrate gezeigt, insbesondere solche mit aktivierten Eosinophilen.

In diesen Arbeiten haben wir Theophyllin über eine Periode von
über 6 Wochen gegeben und uns dabei auf niederen Plasmaspiegeln
bewegt.

Wenn daher Theophyllin solch wichtige Effekte hat, sollten wir
seine Rolle in der Behandlung des Asthmas neuüberdenken, insbeson-

dere seinen Platz im Verhältnis zu anderen Antiasthmamedikamenten, wie sie in nationalen und internationalen Konsensusrichtlinien diskutiert werden.

Introduction

Theophylline is a widely used treatment for asthma, usually in the form of oral sustained release compounds. Despite a recognition of it's therapeutic effectiveness, the mode of action of the xanthine molecules is still not clear either at a cellular or clinical level (Tables 1 and 2). It is now well recognised that there is airway inflammation even in mild asthma [1] and persisting inflammation is a feature of chronic asthma [2]. Therefore anti-inflammatory actions of drugs used in the treatment of chronic asthma assume major importance. It is now thought that theophylline is not just a bronchodilator but, in fact, has significant anti-inflammatory effects [3]. The evidence for this is based on in vitro work, animal and human data. Furthermore, several of these studies have also shown immunomodulatory actions for theophylline.

Table 1. Proposed cellular mechanisms of action of theophylline

---

Inhibition of cyclic nucleotide phosphodiesterases
Interaction with guanine nucleotide regulatory proteins (G Proteins)
Adenosine antagonism
Effects via catacholamine release
Effects on phospholipid metabolism
Effects on Ca^{2+} availability and utilization

Table 2. Proposed clinical effects of theophylline

Bronchial smooth muscle relaxation
Anti-inflammatory effects
Improved mucociliary transport
Increased efficiency of diaphragmatic contraction
Central respiratory stimulation

In vitro studies: effects on cells

Eosinophils

Theophylline and isbufylline inhibit eosinophil recruitment into the airways after bronchial challenge [4, 5]. Also theophylline inhibits eosinophil infiltration in the airway of actively sensitised guinea pigs [4, 6]. A recent study from our own group has shown that theophylline administered in low dose for a period of 5 weeks to atopic asthmatics reduces total eosinophil infiltration in the airway, and in particular reduces the number of eosinophil granule (EG2) positive eosinophils in bronchial biopsies following bronchial challenge [7].

Lymphocytes

Early studies in renal transplant patients showed that theophylline allowed a reduction in the corticosteroids dose required to control rejection and that this was accompanied by a fall in the CD4/CD8 T-cell ratio [8]. As an additional action, theophylline increases T suppressor cell function during renal transplant rejection episodes and amplifies the degree of inhibition of the graft-versus-host response.

Further studies in asthmatics indicate that those taking theophylline have more peripheral blood T-cells that are able to suppress plaque formation in lymphocyte culture than in those taking other anti-asthmatic remedies [9]. When a sample of the blood of these asthmatics is incubated with theophylline it can be shown that it inhibits autologous cell responses, and this can be demonstrated to be due to a specific T-cell sub-group which is sensitive to in vitro stimulation by theophylline [10].

T-suppressor lymphocytes numbers are restored towards normal in the blood of asthmatic children by the administration of theophylline [11, 12].

Mast cells

The rat peritoneal mast cell is a connective tissue type cell similar to those in the lung; theophylline reduces in vitro histamine release from these cells after stimulation by antigen or by anti IgE receptor antibodies [13]. It also inhibits histamine release from human basophils and lung fragments [14].

Monocytes and macrophages

In blood from asthmatic children, theophylline inhibits mononuclear cell chemotaxis ex vivo [15]. Peripheral blood mononuclear cells and alveolar lavage macrophages from normal subjects secrete less superoxide anion after exposure to low concentrations of theophylline in vitro [16]. Theophylline inhibits leukotriene B_4 formation by cultured human monocytes via cyclic adenosine monophosphate and prostaglandin E_2 production [17]; however the biological significance of this is unclear.

Neutrophils

Theophylline reduces the magnitude of the respiratory neutrophil burst after stimulation with either NMLP, a peptide receptor antagonist, or calcium ionophore, with 40% inhibition as therapeutic concentrations [18]. Neutrophil LTB_4 reduction is reduced by 90% by therapeutic concentrations in vivo [19]. Theophylline also reduces neutrophil chemotaxis in the blood of asthmatic children given theophylline for 10 days [16].

Natural killer cells

Although these cells are not thought to be involved in the inflammatory processes in asthma they are relevant to theophylline immunomodulatory activities in general. Theophylline modulates the activity of natural killer cells producing a dose dependent reduction in activity [20]. It is unlikely that this is of significance in relation to tumour development.

In vivo animal studies

The most commonly used model to study the anti-inflammatory effect of theophylline in animals has been antigen induced bronchoconstriction which causes air-flow obstruction and acute airway inflammation. There is some difficulty in interpreting the data as most of the studies have simply used one dose of theophylline and also the doses administered in most reports are much higher than those used clinically. However, theophylline has been shown to be effective in reducing airway inflammation and hyperresponsiveness in the rat model [21, 22] and in the late asthmatic response in the guinea pig [23, 24]. The Gristwood study [24] compared the effects of prednisolone, theophylline and salbutamol in terms of bronchodilator action and inhibition of inflammatory infiltrate in the guinea pig and concluded that prednisolone had effects on leukocyte infiltration but little effects as a bronchodilator. Salbutamol had little effect on either process but theophylline had both bronchodilator and anti-infiltrative actions. Methylxanthines have also been shown to attenuate the late asthmatic response in allergic rabbits [25].

There is one study of theophylline in animal models that has used a dose similar to that administered in man [26]. This demonstrated that theophylline and the more specific phosphodiesterase inhibitor bendzafentrine caused inhibition of pulmonary eosinophil infiltration in the guinea pig airway following platelets activating factor and allergen challenge.

Human studies

Allergic rhinitis has been used as a model to study the anti-inflammatory actions of theophylline. Seven days treatment with theophylline reduces histamine release following antigen challenge in subjects with allergic rhinitis [27]. These authors also showed a reduced skin response to allergen in subjects treated with theophylline. Furthermore, theophylline reduces the plasma nasal exudate in atopic subjects.

The late asthmatic response (LAR) to inhaled allergen is considered to be an in vivo model for inflammation in asthma, representing recruitment and activation of inflammatory cells. There are now a series of studies indicating that theophylline has potent inhibitory effects on the LAR. The first of these studies from Pauwels [28] showed that a single intravenous dose of either theophylline or enprofylline abolished the late asthmatic response to antigen with little effect of the early asthmatic response (EAR). Other investigators have shown variable effects on the LAR [29, 30] but it appears from these data that it is comparable to cromoglycate although probably a less potent inhibitor than inhaled corticosteroids.

In view of the variability of the previously available data we performed a double blind randomized parallel group study comparing the effect of theophylline with placebo on the LAR in 19 asthmatic subjects over a period of 5 weeks. There was no effect of placebo but the

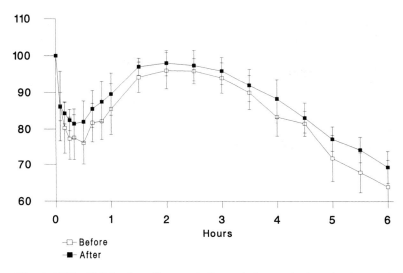

Fig. 1. FEV$_1$ (SEM) after allergen challenge before and after placebo as % baseline

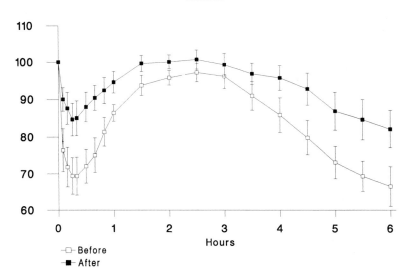

Fig. 2. FEV$_1$ (SEM) after allergen challenge before and after theophylline as % baseline

patients on theophylline had a highly significant attenuation of the LAR, many of them at low plasma levels (mean trough blood theophylline 7.8 mgs/L) [31] with little effect on the EAR. We went on to perform a further study [7] targeting low plasma levels and including bronchial biopsy studies. Once again attenuation of the LAR was demonstrated (Figs. 1, 2) and, as previously quoted, the number of activated eosinophils in the airways of those treated with theophylline was significantly reduced. We concluded from this that low dose oral theophylline attenuates airway inflammation as demonstrated both functionally and histologically in atopic asthmatics.

Conclusion

Although the precise mode of anti-inflammatory action of theophylline is still not certain, it is likely that the effects are being achieved through phosphodiesterase inhibition. We should now apply our new knowledge of these actions of theophylline to maximise the anti-inflammatory benefit for asthmatic subjects. This might include placing children on long-term low dose theophylline, and carries the particular advantage of being an oral medication ensuring better compliance. It is possible that this form of therapy could prevent the structural change in the airway which is seen in untreated chronic asthmatics. If these data are substantiated a review of the position of theophylline in asthma guidelines might be indicated.

References

1. Beasley R, Roach WR, Roberts JA, Holgate ST (1989) Cellular events in the bronchi in mild asthma and after bronchial provocation. Am Rev Respir Dis 139: 806–817
2. Djukanovic R, Roach WR, Wilson JW et al (1990) Mucosal inflammation in asthma. Am Rev Respir 142: 434–457
3. Costello J (1994) The anti-inflammatory actions of theophylline. In: Methylxanthines and phosphodiesterase inhibitors in the treatment of airways disease. Parthenon Press, pp 145–154
4. Sanjar S, Aoki S, Kristersson A, Smith D, Morley J (1990) Antigen challenge induces pulmonary airways eosinophil accumulation and airway hyperreactivity in sensitised guinea pigs; the effect of anti-asthma drugs. Br J Pharmacol 99: 679–686

5. Manzini S, Perretti F, Abelli L, Evangelista S, Seeds EAM, Page CP (1993) Isbufylline, a new xanthine derivative inhibits airways hyper-responsiveness and airway inflammation in guinea pigs. Eur J Pharmacol 249: 251–257

6. Llupia J, Fernandez AG, Berga P, Gristwood RW (1991) Effects of prednisolone, salbutamol and theophylline on bronchial hyperreactivity and leukocyte chemokinesis in guinea pigs. Drugs Exp Clin Res 17: 395–398

7. Sullivan P, Dekir S, Jaffar Z, Page CP, Jeffrey,P, Costello J (1994). Anti inflammatory effects of low dose oral theophylline in atopic asthma. Lancet 343: 1006–1008

8. Guillou PJ, Ramsden C, Kerr M, Davison, AM, Giles GR (1984) A prospective controlled clinical trial of aminophylline as an adjunctive immunosuppressive agent. Transplant Proc 16: 1218–1220

9. Fink G, Mittelman M, Shohat B, Spitzer SA (1987) Theophylline-induced alterations in cellular immunity in asthmatic patients. Clin Allergy 17: 313–316

10. Shore A, Dosch H, Gelfand E (1978) Induction and separation of antigen dependent T helper and T suppressor cells in man. Nature, London 264: 586–587

11. Lahat N, Nir E, Horenstein L, Colin AA (1985) Effect of theophylline on the proportion and function of T-suppressor cells in asthmatic children. Allergy 40: 453–457

12. Shohat B, Volovitz B, Varsano I (1983) Induction of suppressor T cells in asthmatic children by theophylline treatment. Clin Allergy 13: 487–493

13. Pearce FL, Befus AD, Gauldie J, Bienenstock J (1982). Effects of anti-allergic compounds on histamine secretion by isolated mast cells. J Immunol 128: 2481–2486

14. Louis RE, Radermecker MF (1990) Substance P induced histamine release from human basophils, skin and lung fragments: effect of nedocromil sodium and theophylline. Int Arc Allergy Appl Immunol 92: 329–333

15. Condino-Neto A, Vilea MM, Cambiucci EC, Ribeiro JD, Guglielmi AA, Magna LA, De Nucci G (1991) Theophylline therapy inhibits neutrophil and monomuclear cell chemotaxis from chronic asthmatic children. Br J Clin Pharmacol 32: 557–561

16. Calhoun WJ, Stevens CA, Lambert SM (1991) Modulation of superoxide production of alveola macrophages and periferal blood mononuclear cells by beta-agonists and theophylline. J Lab Clin Med 117 (6): 514–522

17. Juergens UR, Overleck A, Velter M (1993) Theophylline inhibits the formation of LTB_4 by enhancement of cAMP and PGE2 production in normal monocytes in vitro. Eur Resp J 6 [Suppl 17]: 3685

18. Neilson CP, Crowley JJ, Cusack BJ, Vestal RE (1986) Therapeutic concentrations of theophylline and enprofylline potentiate catecholamine effects and inhibit leukocyte activation. J Allergy Clin Immunol 78 (4 pt. 1): 660–667

19. Nielson C, Crowley JJ, Morgan M, Vestal R (1988) Polymorphonuclear leukocyte inhibition in therapeutic concentrations is mediated by cyclic AMP. Am Rev Respir Dis 137: 25–30

20. Yokoyana A, Yamashita N, Mizushima Y, Yano S (1990) Inhibitions of natural killer cell activity by oral administration of theophylline. Chest 98: 924–927

21. Kips JC, Pauwels R, Van der Straetan M (1989) Effect of theophylline on the endotoxin-induced airway inflammation and bronchial hyperresponsiveness. J Allergy Clin Immunol 83: 178

22. Pauwels R (1987) The effects of theophylline on airways inflammation. Chest 92 [Suppl]: 32–37S

23. Hutson P, Church M, Clay T, Miller Pt, Holgate ST (1988) Early and late phase bronchoconstriction after allergen challenge of non-anaesthetised guinea pigs. Am Rev Respir Dis 137: 548–557

24. Gristwood R, Llupia J, Fernandez A, Berga P (1991) Effects of theophylline compared with prednisolone on the late phase airway leukocyte infiltration in guinea pigs. Int Arch Allergy Appl Immunol 94: 293–294

25. Ali S, Mustaffa S, Metzger WJ (1992) Modification of allergen induced airway obstruction and bronchial hyperresponsiveness in allergic rabbits by theophylline aerosol. Agents actions 37: 165–173

26. Sanjar S, Aoki S, Boubeckeur K, Chapman ID, Smith D, Kings MA, Morley J (1990) Eosinophil accumulation in pulmonary airways of guinea pigs induced by exposure to an aerosol of platelets activating factor: effect of antiasthmatic drugs. Br J Pharmacol 99: 267–272

27. Naclerio RM, Bartenfelder D, Proud D, Togis AG, Meyers CA, Kagey SA, Norman PS, Lichtenstein LM (1986) Theophylline reduces histamine release during pollen induced rhinitis. J Allergy Clin Immunol 78: 874–876

28. Pauwels, R, Van Renterghem D, Van Der Straeten M, Johannson N, Persson C (1985) The effect of theophylline and enprofylline on allergen induced bronchoconstriction. J Allergy Clin Immunol 76: 583–590

29. Cockroft D, Murdock B, Gore B, O'Byrne P, Manning P (1989) Theophylline does not inhibit allergen-induced increase in airway responsiveness to methacholine. J Allergy Clin Immunol 83: 913–920

30. Hendeles J, Harman E, Huang D, Cooper R, Delafuente J, Blake K, O'Brien R (1991) Attenuation of allergen induced airway hyperreactivity and late response by theophylline. Eur Resp J 4 [Suppl 14]: 4815

31. Ward AJM, McKenniff M, Evans JM, Page CP, Costello J (1993) Theophylline – an immunomodulatory role in asthma? Am Rev Respir Dis 147: 518–523

Correspondence: Dr. J. Costello, King's College, School of Medicine & Dentistry, Department of Thoracic Medicine, Bessemer Road, London SE5 9PJ, U.K.

Discussion

Moderator: Thank you Dr. Costello, for this firework of information and implications. Please comments and discussion.

N.N.: John, in your study in Lancet (Ref. 7), the theophylline levels you mentioned there were trough levels then?

Dr. Costello: No, the first ones were trough levels, the second study was steady state.

N.N.: Steady state taken how long after a dose? Was that 4 hours?

Dr. Costello: Close to peak levels, yes.

N.N.: Thank you.

Dr. Lane: Have you related the pharmacokinetics of an oral or intravenous dose of theophylline to the early and late asthmatic response? I just wonder if the effects on the late asthmatic response could not be secondary to a sustained bronchodilatory effect rather than an anti-inflammatory.

Dr. Costello: If it was a sustained bronchodilator effect, you would see the effect on the early response in a much more dramatic way than we do here. If you compare those 5 or 6 studies and the study that we did, for instance with the study of salmeterol, that was published in the Lancet* about 4 years ago, there was indeed an early and sustained bronchodilator response, but no late response at all. In fact a late response did occur in those subjects, but the functional antagonist effect, the bronchodilator response of the early response was masking it. If the effects we were seeing were bronchodilator, we'd see the effect in a much more dramatic way in the early response. We do achieve a little bronchodilator response but not enough to account for the inhibition of the late response.

N.N.: Could you be missing an early bronchodilator effect purely because you're not achieving sufficient levels in the serum?

Dr. Costello: Yes. It also depends on when you do the challenge test, after the dose of theophylline. But in our second study, we were measuring plasma steady state levels anyway. So it actually doesn't matter.

Dr. Keller: As I understood you, you made the biopsies after the allergen provocation?

Dr. Costello: That's right.

---

* Twentyman OP et al (1990) Lancet 336: 1338.

Dr. Keller: How do you feel about a so-called baseline biopsy study?

Dr. Costello: Well, we thought about that. It's quite a difficult business persuading asthmatics to have two bronchoscopies anyway, and then asking them to have 3 or more. At one time we would have needed 6 to get the perfect study. And it's quite difficult to achieve.

Dr. Lane: The dose of 36 micro Mol was achieved in your Lancet paper. Has that dose been achieved on the cells in the in vitro studies? What was the dose range of in vitro?

Dr. Costello: The dose range of in vitro studies is enormous, as you saw in the studies for instance on beta-adrenergic release and antihistamines, that was several degrees of magnitude greater than that.

Dr. Lane: Are the dose response curves bellshaped or bi-bellshaped?

Dr. Costello: I don't know the answer to that.

Dr. Vignola: Did you see any differences in the epithelial shedding after treatment with theophylline?

Dr. Costello: It wasn't something we quantified in any way. Dr. Jeffery, you saw many of these. The design of the study was such that we couldn't really investigate these biopsies for one reason or another, they were all snap frozen. And in that sort of situation, measuring the extent of shedding is in my opinion an impossibility, because much of the epithelium is lost during that snap freezing procedure.

The effect of theophylline on the chronobiology of inflammation in asthma

G. E. D'Alonzo

Divison of Pulmonary and Critical Care Medicine,
Temple University Medical School, Philadelphia, U.S.A.

Summary

Asthma is an inflammatory disease of the airways. It appears that the inflammatory environment responds differently at night when the asthmatic patient should be asleep as compared to other times during the day. The reason for this difference is multifactorial and likely very complex. Sustained-release theophylline therapy has proven to be an antiasthmatic therapy with impressive efficacy for the control of asthma and nocturnal asthma, the latter especially if administered once-daily in the evening. In that theophylline is a "weak" bronchodilator, it is possible that theophylline works through non-bronchodilator anti-asthmatic mechanisms. Therefore, it may be that theophylline's principal anti-asthmatic effects are antiinflammatory in character. Recent data suggesting that theophylline has certain immunomodulating properties opens a whole new area for clinical investigation with this medication, a therapy that is literally decades old. In that theophylline therapy is relatively low in cost and convenient to take, this new investigative challenge should be taken seriously.

Zusammenfassung

Der Effekt von Theophyllin auf die Chronobiologie der Entzündung beim Asthma. Asthma ist eine entzündliche Erkrankung der Atemwege mit gesteigerter bronchialer Reaktivität. Sowohl die Nor-

malpopulation als auch die Asthmapatienten zeigen die besten Werte
der Atemflüsse um etwa 16 Uhr, die schlechtesten um etwa 4 Uhr
morgens. Die größte Amplitude in der Normalpopulation beträgt etwa
8%, während bei den Asthmatikern dieselbe bis zu 50% beträgt.

Die Intensität dieses Abfalls steht für gewöhnlich zum Schweregrad
der Erkrankung des Individuums in direkter Beziehung. Unsere Kennt-
nisse über die Ausschüttung von Mediatoren beim Asthma stammt
vorwiegend aus Studien während des Tages, während viel weniger über
den Mediatorausstoß und deren Funktion während der Nacht bekannt
ist. Nächtliches Asthma ist derzeit ein wichtiges Modell für die natürli-
cherweise auftretende Entzündung und das Ödem. Ein Wechsel des PEF
über den Tag, Epinephrinspiegel im peripheren Blut, cAMP und Plas-
mahistaminspiegel beeinflussen die bronchiale Konstriktion während
der Nacht. Ein gesteigerter Histaminspiegel spiegelt eine Änderung des
entzündlichen Milieus in der Nacht im Vergleich zum Tag wieder.
Zusätzlich findet man eine signifikante Erhöhung der Entzündungszellen in der BAL-Flüssigkeit bei Patienten mit Asthma, deren Lungenfunktion sich in der Nacht verschlechtert.

Als mögliche Erklärung für diese Veränderungen galt der Grad der
Bronchokonstriktion, das heißt, daß diese erst den Influx der Zellen
auslösen würde. Um dies weiter zu untersuchen, wurde bei 3 Patienten
mit nächtlichem Asthma um 16 Uhr ein Spasmus provoziert, der etwa
den selben Grad erreichen sollte wie jener, der sich spontan um 4 Uhr
Früh zu ereignen pflegte. Die Ergebnisse zeigten, daß es vielmehr die
Zellen selbst sind, die für die Verschlechterung der Lungenfunktion
während des Schlafes primär verantwortlich sind, und daß sie nicht erst
als Folge der Bronchokonstriktion in die Luftwege einwandern.

Die spezifische Rolle der Eosinophilen, Mastzellen und Neutrophilen beim nächtlichen Asthma wird immer deutlicher. Wenn angenommen werden darf, daß dies durch Zytokine geschieht, so ist gerade die T-
Zelle imstande, durch den Ausstoß verschiedener Zytokine die Rekru-
tierung und Aktivierung anderer Entzündungszellen zu steuern. Es
wurde vermutet, daß die Abnahme von zirkulierenden Katecholaminen
die Atemwegsentzündung beim Asthma durch Modulation der T-Zellen
in der Lunge und deren Ausstoß von proinflammatorischen Zytokinen
beeinflußt. Es soll zu der circadianen Veränderung in der Aktivierung
der Eosinophilen, Neutrophilen, Mastzellen und Makrophagen kom-
men, wenn durch den Abfall der Katecholamine und Kortikosteroide

während der Nacht eine gesteigerte Funktion der T-Zellen in der Lunge resultiert. Die pathophysiologischen Folgen sind Bronchospasmus, Ödem und gesteigerte Sputumproduktion.

Bei Patienten – auch bei solchen unter Steroiden – mit schwerem nächtlichen Asthma konnte gezeigt werden, daß höhere Serumtheophyllinkonzentrationen während der Nacht und geringere während des Tages therapeutisch sehr erfolgreich sind.

Die Chronotherapie des Asthmas mit entsprechend langwirkenden Theophyllinpräparaten bewirkt bei einmalig abendlicher Dosis einen Spitzenwert des Blutspiegels während der Nacht und in den frühen Morgenstunden, wenn die Obstruktion bei vielen Patienten mit Asthma am stärksten ist. Diese Medikamente gestatten eine bessere nächtliche Kontrolle des Asthmas mit signifikant besseren Lungenfunktionswerten am Morgen. Zusätzlich scheint die einmalige Gabe am Abend eine günstige Wirkung auch auf die anderen Zeiten während des Tages zu haben.

Wie erwähnt, ist Theophyllin selbst ein schwacher Bronchodilatator, kann aber aufgrund seines Effektes auf die Entzündungsmechanismen, besonders während der Nacht, als „antiasthmatische" Medikation gesehen werden. Die Beobachtung, daß Theophyllin die asthmatische Spätreaktion nach Allergenprovokation beeinflußt, ist hier besonders wichtig, weil Theophyllin möglicherweise vor der Entwicklung des Ödems schützt, ein wichtiger, wenn auch wenig beachteter Mechanismus des nächtlichen Asthmas. Der bronchodilatatorische Effekt des Theophyllins wäre dann ein willkommener Additivfaktor.

Theophyllin steigert außerdem den mukoziliären Transport, vermindert die Ausschüttung von bestimmten Entzündungsmediatoren und unterdrückt das Permeabilitätsödem. Vorbehandlung mit Theophyllin hat eine Verminderung der nasalen Erregbarkeit auf Antigen gezeigt und unterbindet die Ausschüttung von Mediatoren aus Mastzellen. Theophyllin als orale Medikation bei Asthmatikern schützt gegen Allergenprovokation und scheint zur selben Zeit teilweise die Ausschüttung von Histamin zu verhindern.

Zusammenfassend ist die Therapie mit Retard-Präparaten von Theophyllin nachweislich imstande, das nächtliche Asthma erfolgreich unter Kontrolle zu bringen. Asthma ist eine entzündliche Erkrankung, aber Theophyllin nur ein schwacher Bronchodilatator; daher ist des denkbar, das Theophyllin seine Wirkung über andere „antiasthmatische" Mechanismen und nicht nur über die Bronchodilatation erhält. Allerdings ist es

zum gegenwärtigen Zeitpunkt schwierig, Theophyllin als antientzündlichen Stoff schlechthin zu bezeichnen.

Introduction

Asthma is an inflammatory disease of the airways characterized by increased bronchial hyperresponsiveness to provocative challenges with
nonspecific chemical agents and/or specific antigens and is clinically
displayed as episodic or chronic airflow obstruction with a patient expe-
riencing dyspnea, chest congestion, and wheezing. Studies performed in
a laboratory, as well as outpatient self-assessment studies, verify that
airway patency and dyspnea varies as a circadian rhythm [1–4]. In general, the more unstable and severe the asthma, the greater the circadian
variability in airflow and the more reduced the 24-hour mean level of
airway patency compared to normal values. The severity of nocturnal
asthma correlates with the degree of bronchial reactivity [5]. In patients
who are active during the daytime, the airways are least vulnerable to
provocation during the afternoon hours and most susceptible to bronchospasm at night and in the early morning hours during sleep [6–11].

The normal population and asthmatic population display best airflow
function at approximately 4:00 p.m. and their nadir at 4:00 a.m. The peak-
to-trough difference in airflow in the normal population is approximately
5% to 8%, whereas the asthmatic population can have substantially higher peak-to-trough swings, at times as high as 50% [4, 12]. Although the
mechanism for this 24-hour variability remains unknown, it is likely that
numerous interplaying factors occur and cause airflow to fall in the early
morning hours of each day in most patients with asthma [13]. The intensity of the fall in airflow that can occur relates to the degree of disease that
an individual patient has. Generally speaking, the greater the overnight
fall in peak expiratory flow rates, the larger the circadian change in bronchial reactivity. Even those asthmatics without nocturnal symptoms have
a significant increase in airway reactivity at night.

Pathogenesis of nocturnal asthma

In an effort to explain why asthma seems to have an expressive predilec-
tion for the nighttime, a number of hypotheses have been proposed,
including day-night differences, environmental factors, such as baro-

metric pressure, relative humidity and ambient temperature, proximity and concentration of various offending antigens, accumulative effects of psychological and physiological stresses during the day, and supine posture at night [13]. The airways in both normal and asthmatic subjects narrow during sleep; and the extent of overnight airway narrowing is reduced when asthmatic subjects are kept awake overnight [14]. Some nocturnal airway narrowing persists during a single night of sleep deprivation and thus sleep alone does not directly cause all of the nocturnal airway narrowing. Functional residual capacity falls during sleep [15, 16]. This change will passively narrow the bronchial tree. Recent evidence suggest that the reduction in FRC in patients with asthma may result in bronchoconstriction that persists after awakening and thus will contribute to the overnight airway narrowing seen in patients with nocturnal asthma [16, 17]. Additionally, there is evidence that some patients with asthma, who snore loudly, develop overnight airway narrowing as a direct consequence of their snoring [18, 19].

Chronobiologic (biological rhythm) studies, provide an alternative explanation for this disease pattern by stressing the role of endogenous circadian bioperiodicities in relationship to changes in the external environment that occur over each 24 hour [13]. Parasympathetic tone increases during the night, whereas sympathetic influence decreases. In patients with nocturnal asthma, increased parasympathetic tone may be a major factor in the pathogenesis of its expression. The sympathetic nervous system does not directly innervate the bronchial smooth muscle, however, the reduction in plasma epinephrine likely plays an important consequence in nocturnal asthma. The only other direct intervention of the bronchial smooth muscle is the non-adrenergic, noncholinergic system for which nitric oxide is the likely neurotransmitter. Recent evidence indicates that this particular part of the autonomic nervous system is inhibited in the early morning and this could enhance bronchoconstriction [20].

Nocturnal asthma and airway inflammation

As stressed, asthma is an inflammatory disease of the airways. Careful histopathological studies of airways from the lungs of patients who have died from asthma have highlighted the importance of airway inflammation, comprising of mast cell degranulation, infiltration by eosinophils and mononuclear cells, epithelial disruption, hypertrophy of airway smooth muscle, hypersecretion of mucous, increased tissue microvasculature leakage and edema formation.

Until recently, our knowledge of the inflammatory process came from studies conducted mainly during the daytime hours and little was known about mediator release and inflammation at night. Nocturnal asthma is actually an important model for naturally occurring, nonchallenged inflammation and edema of the airways. Circadian changes in peak expiratory flow, plasma epinephrine, cyclic-AMP, and cortisol favor bronchoconstriction at night [13, 21, 22]. The suggestion that plasma histamine concentrations rise at night and contribute to the development of nocturnal asthma is based on a study [21] with methodological concerns [22]. A more recent study evaluating a larger number of patients with nocturnal asthma, showed no rise in plasma histamine concentration despite a substantial fall in overnight peak flow rate [23]. Nonetheless, other studies implicate a change in the inflammatory environment at night as compared to the daytime.

Considerable attention has been focused on a more direct assessment of airway inflammation by bronchoalveolar lavage (BAL) in patients with nocturnal asthma. Martin and colleagues reported a significant increase in BAL fluid inflammatory cells in patients with asthma whose lung function is worse at night [24]. An increase in total white cells, neutrophils, and eosinophils in lavage fluid is found at 4 a.m. in comparison with that obtained at 4 p.m. in patients with nocturnal asthma. MacKay et al., similarly found significant increases in eosinophil numbers in lavage fluids at 4 a.m. in patients with nocturnal asthma and also found an increase in lymphocyte numbers, however no change in total white cell count or neutrophils were noted [25]. MacKay et al. not only found an increase in eosinophil numbers in lavage fluid, but also an increase in eosinophil cationic protein in lavage fluid at 4 a.m. in patients with nocturnal asthma. Jarjour et al., failed to show an increase in cell numbers in lavage fluid between the 2 times in patients with nocturnal asthma, though only 5 patients were evaluated, but they did report an early morning increase in superoxide production from lavage cells in their patients with nocturnal asthma [26]. Neither MacKay nor Jarjour's group found any change in histamine concentration in the lavage fluid at 4 a.m.

These studies suggest that inflammatory activity in the airways is increased in the early morning in patients with nocturnal asthma. The trigger for this hyperactivity is not clear. Bronchoconstriction could have precipitated the influx of inflammatory cells. To investigate this possibility further, bronchial constriction was induced in 3 patients with nocturnal asthma at 4 p.m. to the same degree as it occurred at 4 a.m. and failed to show a change in the inflammatory picture by BAL [24]. Since plasma epinephrine falls in the early morning hours, it is tempting to invoke that the circadian rhythm in this catechol is responsible, at least in part, for the deterioration in asthma that occurs overnight. Studies have attempted to replace plasma epinephrine levels to daytime levels overnight and have shown that in certain patients nocturnal asthma is ameliorated while in others it persists [27, 28].

In diurnally active subjects, the highest plasma cortisol levels occur around the time of awakening from sleep in the morning and the lowest levels occur in the middle in the night [21, 29, 30]. The peak-trough variation in this rhythm is generally very large. Since cortisol can potentially effect airway function in asthmatic patients, it was initially proposed that differences in airway patency between asthmatic and non-asthmatic individuals could be due to disparities in cortisol secretion. Investigators have attempted to establish a relationship between circadian variations in airway caliber and plasma cortisol. Reinberg et al. demonstrated synchrony in the timing of nocturnal bronchoconstriction and the lowest urinary excretion of 17-hydroxycorticosteroid over a 24-hour period [31]. Barnes et al. found that the nadir of plasma cortisol occurred at midnight, whereas the maximum airflow fall that occurred at night was generally at 4 a.m. in a small group of asthmatics [21]. Since the airways of asthmatics tend to be chronically inflamed, temporal changes in plasma cortisol, which likely exert an antiinflammatory effect, may contribute to the day-night pattern in airway patency and the risk for asthma. Souter et al. infused variable physiologic doses of hydrocortisone in a group of 6 patients with nocturnal asthma [30]. Physiologic doses of corticosteroid did not totally block the nocturnal fall in peak expiratory flow rate in 5 of 6 subjects. Thus, with physiologic doses administered acutely, improvement in asthma was not seen. In an attempt to override the nocturnal inflammatory effect, Beam and colleagues used large doses of corticosteroid therapy [32]. In 9 of the 11 nocturnal asthmatics studied, supraphysiological corticosteroid infusion during sleep resulted in greater than 40% improvement in the overnight fall in FEV_1. Other studies, which used corticosteroid chronotherapy acutely and chronically, were also able to attenuate the deterioration in asthma seen at night in asthmatic patients [33–35]. The ability of high or chronic dosing with corticosteroid to significantly attenuate nocturnal asthma points to the role of inflammation in the pathogenesis of this common clinical expression of asthma.

The specific roles of eosinophils, mast cells and neutrophils in nocturnal asthma have begun to emerge. Other cells like lymphocytes and macrophages are also potentially important in this process and, no doubt, interact with other inflammatory cells. While a definite conclusion as to the course of nocturnal asthma cannot be made from available data, a plausible hypothesis has evolved that centers on activation and recruitment of eosinophils, neutrophils, and mast cells to the airways [36]. From emerging evidence, this process is most likely influenced by a variety of cytokines. Since the T-cell is capable, through the release of cytokines, of recruiting and activating other inflammatory cells including the mast cell, eosinophil, macrophage and neutrophil and because of the known inhibitory effects of catecholamines on T-cells, it has been proposed that changes in the levels of certain circulating hormones influence airway inflammation in asthma by modulating pulmonary T-cell release of pro-inflammatory cytokines, thus causing a circadian change in the function of these inflammatory cells. Decreased levels of catecholamines and steroids at night allow for enhanced pulmonary T-cell function which, in turn, leads to activation of airway inflammatory cells at night. The pathophysiologic effect of these changes translates into the bronchospasm, airway edema and increased sputum production.

Bronchial hyperreactivity and nocturnal asthma

As mentioned previously, the severity of nocturnal asthma correlates with the degree of bronchial reactivity. Those patients with a greater fall in airflow at night have an increased bronchoconstrictor response to histamine or methacholine during the daytime [5, 6]. Additionally, nocturnal worsening of asthma is dependent on a circadian rhythmicity in airway hyperreactivity [7–10]. Asthmatic patients, when challenged by the inhalation of histamine or methacholine at different clock hours of the day and night, exhibit a large variation in the threshold dose required to produce a 15 to 20% reduction in bronchial patency. In diurnally active patients, the airways are more reactive when challenged in the early morning, when asthma patients are typically asleep then later in the day, while awake. Similar results have been found when house mite dust challenge was performed in selected sensitive asthmatics [9].

For all provoking agents thus far studied, except for bronchial challenge with cold air and eucapnic hyperventilation, hyperreactivity of the airways during the nighttime is far greater than during the daytime. The acute asthmatic reaction that occurs after exposure to allergen is likely to be more easily provoked in the middle of the night as compared to any other time period during the day. To further assess the time relationship of antigen provocation to intensity of airflow obstruction and particularly the late asthmatic reaction, Mohiuddin and Martin [10] compared the effect of morning and evening antigen challenge on the frequency, time of onset, and severity of the induced late-asthmatic action. In this placebo-controlled study, an antigen challenge performed at 8 a.m. was less likely to induce a late-asthmatic reaction when compared to an antigen challenge occurring at 6 p.m. Moreover, the antigen challenge that occurred in the late afternoon caused a greater fall in FEV_1 during the late-asthmatic reaction and the duration of this fall in FEV_1 was prolonged. Finally, airway hyperresponsiveness postantigen challenge was significantly greater at 24 hours following the evening antigen challenge as compared to the morning challenge. This observation suggests that factors determined by normal circadian rhythms likely promote and accelerate the development of the late-asthmatic reaction to antigen exposure.

Airway responsiveness can be measured using different stimuli. Challenges with methacholine and histamine can invoke airway narrowing mainly by a direct action on receptors on airway smooth muscle. Stimulation with AMP and propranolol are thought to act on other cells which initiate processes that indirectly lead to smooth muscle contraction [37, 38]. Although the exact mechanism leading to airway obstruction induced by AMP inhalation remains unclear, it is likely that this agent is rapidly metabolized to adenosine which acts by stimulating purinoceptors on cell surfaces [38]. Different investigations indicate that a direct effect of adenosine on smooth muscle is unlikely and it is generally thought that this agent stimulates mast cells which when activated indirectly induce bronchospasm. If inflammatory processes are involved in the nocturnal increase in airway responsiveness, then assessment of indirect rather than direct airway responsiveness may better reflect the status of the mucosal environment in-vivo. Osterhof et al. [39] demonstrated that the circadian change in airway hyperresponsiveness to AMP is significantly related to the circadian peak expiratory flow rate variation. Moreover, daytime AMP-induced bronchial reactivity was significantly higher in the group with increased circadian peak expiratory flow rate amplitude than in a group of asthmatics with a much lower degree of 24-hour airflow variability. They suggested that the higher susceptibility to stimulation of indirect airway responsiveness suggests that mast cell activation rather than changes in smooth muscle contractility play a role in the development of nocturnal asthma.

Theophylline therapy

Theophylline has been used for over 50 years in the management of asthma and on a worldwide basis, it is one of the most widely used antiasthma drug. In recent years, however, its role has been challenged [40–42], certainly by the emphasis on early institution of inhaled antiinflammatory medication and, more recently, with the advent of long-acting inhaled β_2-adrenergic agonists [41, 42]. Theophylline is considered a weak bronchodilator when compared to β_2-agonist therapy. However, there are data that strongly argue that theophylline therapy has substantial value as maintenance therapy for patients with chronic disease including those with nocturnal asthma [43–48] and those with systemic and inhaled topical steroid dependency [44, 49, 50].

Mechanisms of action

Several mechanisms of action have been proposed as being responsible for theophylline's beneficial effects in asthma [51–53]. It could be that more than one mechanism are operative. Theophylline has been proposed to inhibit phosphodiesterase, antagonize the adenosine receptor, increase circulating epinephrine, inhibit calcium ion flux and reduce mediator release from inflammatory cells.

Currently, it is generally accepted that theophylline's principle mechanistic effect is inhibition of phosphodiesterase, which leads to an increase in intracellular cyclicnucieotide concentration [51, 52]. Theophylline is a non-selective phosphodiesterase inhibitor and it is generally considered to be a weak one, at best [54, 55]. Some phosphodiesterases are more important for either the regulation of inflammatory cell regulation or for smooth muscle relaxation than others, but there is no evidence that theophylline has a greater inhibitory effect on any one isoenzyme [54, 55]. However, it is possible that certain phosphodiesterase isoenzymes may have an increased expression in the inflammatory environment of the asthmatics airways. Alveolar macrophages harvested from asthmatics' airways have an increased phosphodiesterase activity [56, 57] and theophylline may have a greater than expected inhibitory effect in this setting. This would mean that theophylline may have a greater inhibitory effect on phosphodiesterase in asthmatic airways than in normal airways.

The ability of theophylline to reduce bronchial hyperresponsiveness has been questioned [58–60]; however, there are convincing data which suggest that theophylline reduces the bronchoconstrictive challenge effects of histamine and methacholine [61–68]. In fact, suppression of histamine-induced reactivity is most pronounced when the histamine challenge test performed in the early morning after an evening administration of a sustained-release theophylline [67]. Despite theophylline's variable ability to alter bronchial reactivity, several groups have shown significant suppression of the late-phase response to antigen at modest serum levels of theophylline, substantially more than its inhibition of the immediate-phase reaction [59, 65, 66, 68–72]. These effects of theophylline have been seen at serum theophylline concentrations less than 10 mg/L [70, 71]. If the late-phase response is considered to be a model of airway inflammation of asthma, this would support an antiinflammatory action for theophylline.

Inhibition of phosphodiesterase III (cyclic-GMP inhibited) and IV (cyclic-AMP specific) enzymes by various specific inhibitors produces smooth muscle relaxation, and specific inhibitors of phosphodiesterase IV in the particulate fraction of granulocytes and probably other cells, inhibits their activation and eventual release of proinflammatory mediators [55]. While theophylline is a nonspecific phosphodiesterase inhibitor, it has sufficient inhibitory activity at concentrations in the therapeutic range against extracts of these separated enzymes to be taken seriously when considering mechanisms of both muscle relaxation and suppression of inflammation, especially in the presence of beta-agonists which may involve a synergistic interaction [73].

Recently, as mentioned previously, it has been observed that theophylline at low serum concentrations inhibits the late-asthmatic reaction following allergen challenge, but also suppresses the allergen-induced increase in CD4+ and CD8+ lymphocytes in peripheral blood [70]. In a similarly designed study, Sullivan et al. performed bronchial biopsies 24 hours following allergen inhalation before and after six weeks of theophylline therapy were the serum drug concentrations were below therapeutic levels [71]. After treatment, there was a significant reduction in activated and total eosinophils beneath the airway epithelial basement membrane. Finally the withdrawal of theophylline therapy in patients also treated with high-dose steroids has been associated with deterioration in asthma and an alteration in inflammatory cells in the blood [74, 75]. A fall in lung function paralleled a fall in both activated CD4+ and CD8+ lymphocytes in the peripheral blood, implying that theophylline, at subtherapeutic doses, increases the portion of activated T-lymphocytes in peripheral blood [74]. Bronchial biopsies showed a mirror image of the peripheral blood, with an increase in both CD4+ and CD8+ T-lymphocytes upon theophylline withdrawal [75]. Therefore, theophylline may have an immunomodulating effect involved in transmembrane trafficking of T-lymphocytes from the circulation into the airways.

There are additional data which supports the antiinflammatory effect of theophylline. In patients with allergic rhinitis, theophylline has been shown to block the clinical response to antigen challenge and to decrease the release of certain inflammatory mediators from mast cells [76, 77]. This suggests that theophylline plays a role in the reduction of mast cell and basophil activation. Other investigators have proposed that xanthines may be active at nasal and other airway endothelial-epithelial membranes to reduce leakage of proteinaceous plasma [78–80]. Also, theophylline can attenuate inflammatory stimulus-induced plasma exudation into the guinea pig tracheal wall and lumen [81–83]. Finally, theophylline, at normal concentrations, strongly inhibits the release of sensory neuropeptides in a guinea pig model [84].

Abundant in-vitro (isolated cells or tissue fragments) data exists, but it's relevance to asthma in humans remains to be further defined. Theophylline inhibits histamine release from basophils and mast cells, inhibits adenosine-induced mediator release from mast cells, decreases cytokine release from activated T-lymphocytes, reduces the release of reactive oxygen species from activated macrophages and neutrophils, and decreases the release of major basic protein from eosinophils [85].

Theophylline therapy and nocturnal asthma

There are data that suggest that theophylline has value as a form of maintenance therapy for patients with asthma and in particular, nocturnal asthma. Joad and colleagues studied the benefit of maintenance therapy with theophylline, inhaled albuterol and a combination of these therapies in patients with chronic asthma over a 3 month period [44]. Twice-daily sustained-release theophylline alone or in combination with albuterol was associated with significantly fewer days with symptoms in comparison to treatment with albuterol alone. The greater frequency of symptoms during the albuterol regime was apparent more than 4 hours after dosing and was greatest between 4 and 8 a.m. Several studies confirmed that the use of theophylline provides important additional benefits, particularly in those patients who are already treated with systemic doses of corticosteroid and/or beta agonists [44, 45, 49, 50]. In patients with moderate to severe nocturnal asthma, higher serum theophylline concentrations during the night and lower serum theophylline concentration during the daytime, when there may be less or easier to control bronchial constriction, has been shown to be successful [46–48].

This large body of clinical investigation indicates that the chronotherapy of asthma, with appropriately designed once in the evening sustained-release theophylline medication, can deliver peak theophylline blood levels during the nighttime or early morning hours when airflow obstruction in many patients with asthma is likely to be most severe. The use of these medications affords better nocturnal control of asthma with significantly better early morning airflow than observed with the same daily doses of twice-daily sustained-release theophylline therapy or once-daily in the morning regimes. Additionally but very importantly, once-daily evening therapy provided similar protection at other times during the day.

As mentioned, theophylline is thought to be a weak bronchodilator. Considering this, it is important to attempt to understand how theophylline is effective in stabilizing and preventing nocturnal asthma. In that asthma is an inflammatory disease and there exist a possibility that the inflammatory environment of the airways acts differently at night as compared to the daytime, it is appropriate to explore the possibility of theophylline being an anti-asthmatic drug, not solely a bronchodilator. The observation that theophylline prevents the late-asthmatic reaction to allergen may be particularly relevant because it implies that theophylline provides some protection against the development of airway edema, which may prove to be an important but neglected mechanism of nocturnal asthma. Additional non-bronchodilator effects of theophylline have been proposed to be a therapeutic importance in asthma. As mentioned, pretreatment with theophylline has been shown to reduce the response to nasal challenge with antigen and inhibit the release of mediators from a variety of inflammatory cells. Theophylline, when administered orally to asthmatics, protects against allergen provocation and, at the same time, partially inhibits the release of histamine. Additionally, theophylline has been shown to decrease permeability edema induced by different asthma mediators in the microcirculation of animals.

To date, published data suggests that theophylline alters the inflammatory environment of the airways in patients with nocturnal asthma. However, direct data supporting this hypothesis are needed. Recently, Torvik et al. [86] studied the effect of a once-daily in the evening theophylline preparation on 4 a.m. BAL neutrophil count and function. Eight nocturnal asthmatics were studied in a well designed, placebo-controlled fashion. With evening theophylline therapy, the fall in FEV_1 that occurred overnight was approximately 10% as compared to nearly 27% with placebo administration. A small, but significant reduction in neutrophil count of the BAL fluid was seen with theophylline therapy. It was suggested that the improvement in overnight lung function in nocturnal asthma by theophylline may be related to an alteration of airway neutrophil number and perhaps function. Importantly, neutrophil function data were not presented. More intense investigation on how theophylline therapy alters the airway inflammatory environment at night as compared to other times during the day are needed.

References

1. Reinberg A, Ghata J, Sidi E (1963) Nocturnal asthma attacks; their relationship to the circadian adrenal cycle. J Allergy 34: 323–330
2. Mascia M (1968) Evaluation of night coughing in asthmatic children. J Asthma Dis 5: 163–169
3. Dethlefsen U, Repges R (1985) Ein neues Therapieprinzip bei nächtlichem Asthma. Med Klin 80: 44–47

4. Smolensky MH, Barnes PJ, Reinberg A, McGovern JP (1986) Chronobiology and asthma. Day-night differences in bronchial patency and dyspnea and circadian rhythm dependencies. J Asthma 23: 321–343

5. Ryan G, Latimer KM, Dolovich J, Hargreave FE (1982) Bronchial responsiveness to histamine: relationship to diurnal variation of peak flow rate, improvement after bronchodilator, and airway calibre. Thorax 37: 423–429

6. DeVries K, Goei JT, Booy-Noord H, Orie NG (1962) Changes during 24 hours in the lung function and histamine hyperreactifvity of the bronchial tree in asthmatic and bronchitic patients. Int Arch Allergy 20: 93–101

7. Tammeling GJ, DeVries K, Kruyt EW (1977) The circadian pattern of the bronchial reactivity to histamine in healthy subjects and patients with obstructive lung disease. In: Smolensky MH, Reinberg A, McGovern JP (eds) Chronobiology in allergy and immunology. CC Thomas, Springfield Ill, pp 139–150

8. Van Aalderen WMC, Postma DS, Loëter GH, Gerritsen J, Knol K (1987) Increase in airway hyperreactivity during the night in asthmatic children. Respir Dis 135 (4, Part 2): 460

9. Gervais P, Reinberg A, Gervais C, Smolensky MH, De France O (1977) Twenty-four-hour rhythm in the bronchial reactifvity to house dust in asthmatics. J Allergy Clin Immunol 59: 207–213

10. Mohiuddin AA, Martin RJ (1990) Circadian basis of the late-asthmatic response. Am Rev Respir Dis 142: 1153–1157

11. Martin RJ, Cicutto LC, Ballard RD (1990) Factors related to the nocturnal worsening of asthma. Am Rev Respir Dis 141: 33–38

12. Hetzel MR, Clark TJH (1980) Comparison of normal and asthmatic circadian rhythms in peak expiratory flow rate. Thorax 35: 732–738

13. D'Alonzo GE, Smolensky MH (1991) Chronophysiologic determination of asthma. Ann NY Acad Sci 618: 123–139

14. Ballard RD, Saathoff MC, Patel DK, Kelly PL, Martin RJ (1989) Effect of sleep on nocturnal bronchoconstriction and ventilatory patterns in asthmatics. J Appl Physiol 67: 243–249

15. Hudgel DW, Martin RJ, Johnson B, Hill P (1984) Mechanics of the respiratory system and breathing pattern during sleep in normal humans. J Appl Physiol 56: 133–137

16. Ballard RD, Irvine CJ, Martin RJ, Pack J, Pandey R, White DP (1990) Influence of sleep on lung volume in asthmatic patients and normal subjects. J Appl Physiol 68: 2034–2041

17. Ballard RD, Pack J, White DP (1991) Influence of posture and sustained loss of lung volume on pulmonary function in awake asthmatic subjects. Am Rev Respir Dis 144: 499–503

18. Chan CS, Woolcock AJ, Sullivan CE (1988) Nocturnal asthma: role of snoring in obstructive sleep apnea. Am Rev Respir Dis 137: 1502–1504

19. Guilleminault C, Quera-Salva MA, Powell N, Riley R, Romaker A, Partinen M, Baldwin R, Nino-Murcia G (1988) Nocturnal asthma: snoring, small pharynx, and nasal CPAP. Eur Respir J 1: 902–907

20. MacKay TW, Fitzpatrick MF, Douglas NJ (1991) Non-adrenergic, non-cholinergic nervous system and overnight airway calibre in asthmatics and normal subjects. Lancet 338: 1289–1292

21. Barnes P, Fitzgerald G, Brown M, Dollery C (1980) Nocturnal asthma and changes in circulating epinephrine, histamine and cortisol. N Engl J Med 303: 263–267

22. Mikuni M, Saito Y, Koyama T, Daiguji M, Yamashita I Yamazaki K, Honma A, Ui M (1978) Circadian variation in plasma 3': 5'-cyclic adenosine monophosphate and 3': 5'-cyclic guanosine monophosphate in normal adults. Life Sci 22: 667–671

23. Fitzpatrick MF, MacKay T, Walters C, Tai PC, Church MK, Holgate ST et al (1992) Circulating histamine and eosinophil cationic protein levels in nocturnal asthma. Clin Sci 83: 227–232

24. Martin RJ, Cicutto LC, Smith HR, Ballard RD, Szefler SJ (1991) Airway inflammation in nocturnal asthma. Am Rev Respir Dis 143: 351–357

25. MacKay TW, Wallace WAH, Howie SEM, Brown PH, Greening AP, Church MK, Douglas NJ (1994) Role of inflammation in nocturnal asthma. Thorax 49: 257–262

26. Jarjour NN, Busse WW, Calhoun WJ (1992) Enhanced production of oxygen radicals in nocturnal asthma. Am Rev Respir Dis 146: 905–911

27. Soutar CA, Carruthers M, Pickering CAC (1979) Nocturnal asthma and urinary adrenaline in noradrenaline concentrations. Thorax 32: 677–683

28. Barnes PJ, Fitzgerald GA, Dollery CT (1982) Circadian variation in adrenergic responses in asthmatic subjects. Clin Sci 62: 349–354

29. Postma DS, Keyzer JJ, Loeter GH, Sluiter HJ, DeVries K (1985) Influence of the parasympathic and sympathetic nervous system on nocturnal bronchial obstruction. Clin Sci 69: 251–258

30. Soutar CA, Costello G, Ijaduola O, Turner-Warwick M (1975) Nocturnal and morning asthma: relationship to plasma corticosteroids and response to cortisol infusion. Thorax 30: 436–440

31. Reinberg A, Ghata J, Sidi E (1963) Nocturnal asthma attacks: their relationship to the circadian cycle. J Allergy 34: 323–330

32. Beam WR, Ballard RD, Martin RJ (1992) Spectrum of corticosteroid sensitivity in nocturnal asthma. Am Rev Respir Dis 145: 1082–1086

33. Reinberg A, Gervas P, Choussade M, Fraboulet G, Duburgue B (1983) Circadian changes in effectiveness of corticosteroids in eight patients with allergic asthma. J Allergy Clin Immunol 71: 425–433

34. Reinberg A, Halberg F, Falliers CJ (1974) Circadian timing of methylprednisolone effects in asthmatic boys. Chronobiologia 1: 333–347

35. Beam WR, Weiner DE, Martin RJ (1992) Timing of prednisone and alterations or airway inflammation in nocturnal asthma. Am Rev Respir Dis 146: 1524–1530

36. Jarjour NN, Sheth KK, Busse WW (1993) Cellular mechanisms of nocturnal asthma: the role of the eosinophil neutrophil, and mast cell. In: Martin RJ (ed) Nocturnal asthma: mechanisms and treatment. Futura Publ Co, Mt Kisco, New York, pp 163–197

37. Pauwels R, Joos G, Vender Straeten M (1988) Bronchial hyperresponsiveness is not bronchial asthma. Clin Allergy 18: 317–321
38. Ng WH, Polosa R, Church MK (1990) Adenosine bronchoconstriction in asthma: investigations into its possible mechanisms of action. J Clin Pharmacol 30: 89–98
39. Oosterhoff Y, Koëter GH, De Monchy JGR, Postma DS (1993) Circadian variation in airway responsiveness to methacholine, propranolol, and AMP in atopic asthmatic subjects. Am Rev Respir Dis 147: 512–517
40. Lam A, Newhouse MT (1990) Management of asthma and chronic airflow limitation. Are methylxanthines obsolete? Chest 98: 44–52
41. Newhouse MT (1990) Is theophylline obsolete? Chest 98: 1–2
42. Jenne JW (1990) Theophylline is no more obsolete than "two puffs qid" of current beta-$_2$ agonists. Chest 98: 3–4
43. D'Alonzo GE, Smolensky MH (1993) Chronopharmacology of theophylline and beta-$_2$ sympathomimetic therapies. In: Martin RJ (ed) Nocturnal asthma: mechanisms and treatment. Futura Publ Co, Mt Kisco New York, pp 221–280
44. Joad JP, Ahrens RC, Lindgren SD Weinberger MM (1987) Relative efficacy of maintenance therapy with theophylline, inhaled albuterol, and the combination for chronic asthma. J Allergy Clin Immunol 79: 78–85
45. Zwillich CW, Neagley SR, Cicculto L et al (1989) Nocturnal asthma therapy: inhaled bitolterol versus sustained-release theophylline. Am Rev Respir Dis 139: 470–474
46. Neuenkirchen H, Wilkens JH, Oellerich M, Sybrecht GW (1985) Nocturnal asthma: effect of a once per evening dose of sustained-release theophylline. Eur J Respir Dis 66: 196–204
47. Martin RJ, Ciculto LC, Ballard Rd et al (1989) Circadian variations in theophylline concentrations and the treatment of nocturnal asthma. Am Rev Respir Dis 139: 475–478
48. D'Alonzo GE, Smolensky MH, Feldman S et al (1990) Twenty-four hour lung function in adult patients with asthma: chronoptimized theophylline therapy once-daily dosing in evening versus conventional twice-daily dosing. Am Rev Respir Dis 142: 84–90
49. Nassif EG, Weinberger M, Thompson R, Huntley W (1981) The value of maintenance theophylline in steroid-dependent asthma. N Engl J Med 304: 71–75
50. Brenner M, Berkowitz R, Marshall N, Strunk RC (1988) Need for theophylline in severe steroid-requiring asthmatics. Clin Allergy 18:143–150
51. Barnes PJ, Pauwels RA (1994) Theophylline in the management of asthma: time for reappraisal? Eur Respir J 7: 579–591
52. Greening A (1984) The role of theophylline in the management of asthma. In: Turner-Warwick M, Levy J (eds) New perspectives of theophylline therapy. Royal Society Med London, pp 13–22
53. Pauwels R, Van Der Straeten (1984) The anti-allergic effect of theophylline. In: Jonkman JHG, Jenne JW, Simons FER (eds) Sustained realse theophylline, excepta medica, Elsevier Science Publ Co, Amsterdam, pp 9–13

54. Barnes PJ (1994) Cyclic mulcotide and phosphodiesterase inhibitors. In: Phosphodiesterase – a key enzyme m regulations of smooth muscle contraction and inflammation. Byk Gulden Monograph, Konstanz, Germany

55. Torphy TJ, Undem BJ (1991) Phosphodiesterase inhibitors: new opportunities for the treatment of asthma. Thorax 46: 512–523

56. Baehelet M, Vincent D, Havet N (1991) Reduced responsiveness of adenylate cyclose in alveolar macrophages from patients with asthma. J Allergy Clin Immunol 88: 322–328

57. Townley RG (1993) Elevated camp-phosphodiesterase in atopic disease: cause or effect? J Lab Clin Med 121: 44–51

58. Crockcroft DW, Murdock KY, Gore BP, O'Byrne PM, Manning P (1989) Theophylline does not inhibit allergen-induced increase in airway responsiveness to methacoline. J Allergy Clin Immunol 83: 913–920

59. Mapp C, Boshetto P, Dal Vecchio L, Crescioli S, de Marzo N, Paleari D, Fabbri LM (1987) Protective effect of antiasthma drugs on late asthmatic reactions and increased airway responsiveness induced by toluene diisocyanate in sensitized subjects. Am Rev Respir Dis 136: 1403–1407

60. Dutoit JI, Salome CM, Woolcock AJ (1987) Inhaled corticosteroids reduce the severity of bronchial hyperresponsiveness in asthma but oral theophylline does not. Am Rev Respir Dis 137: 1174–1178

61. McWilliams BC, Menendez R, Kelly WH, Howick J (1984) Effects of theophylline on inhaled methacholine and histamine in asthmatic children. Am Rev Respir Dis 130: 193–197

62. Cartier A, Lemire I, L'Archeveque J, Ghezzo H, Martin R, Malo JL (1986) Theophylline partially ingibits bronchoconstrition caused by inhaled histamine in subjects with asthma. J Allergy Clin Immunol 77: 570–575

63. Crimi N, Palermo F, Distefano SM (1987) Relationship of serum theophylline concentrations to histamine-induced bronchospasm. Respiration 52: 189–194

64. Magnussen H, Reuss G, Jorres R (1987) Theophylline has a dose-related effect on the airway response to inhaled histamine and methacholine in asthmatics. Am Rev Respir Dis 136: 1163–1167

65. Crescioli S, Spinazzi A, Plebani M, Pozzani M, Mapp CE, Boschetto P, Fabbri LM (1991) Theophylline inhibits early and late asthmatic reactions induced by allergens in asthmatic subjects. Ann Allergy 66: 245–251

66. Hendeles L, Harman E, Huang D et al (1991) Attenuation of allergen-induced airway hyperreactivity and late response by theophylline. Eur Respir J 4: P1057

67. Milgrom H, Barnhart A, Gaddy J, Bush RK, Busse WW (1990) The effect of a 24-hour sustained release theophylline (Uniphyl) on diurnal variations in airway responsiveness. J Allergy Clin Immunol 85: 144

68. Koëter GH, Kraan J, Boorsma M, Jonkman JHG, Vender Mark THW (1989) Effect of theophylline and enprofylline or bronchial hyperresponsiveness. Thorax 44: 1022–1026

69. Pauwels R, Van Renterghem D, Vender Straeten M, Johnanesson N, Persson

CGA (1985) The effect of theophylline and enprofylline on allergen-induced bronchoconstriction. J Allergy Clin Immunol 76: 583–590

70. Ward AJM, McKenniff M, Evans JM, Page CP, Costello JF (1993) Theophylline – an immunomodulatory role in asthma? Am Rev Respir Dis 147: 518–523

71. Sullivan P, Bekir S, Jaffar Z, Page C, Jeffery P, Costello J (1994) Antiinflammatory effects of low-dose oral theophylline in atopic asthma. Lancet 343: 1006–1008

72. Cresioli S, de Marzo N, Boschetto P et al (1992) Theophylline inhibits late asthmatic reactions induced by toluene diisocyanate in sensitized subjects. Eur J Pharmacol Environ Toxicol 228: 45–50

73. Neilson CP, Crowley JJ, Morgan ME, Vestal RE (1988) Polymorphonuclear leukocyte inhibition by therapeutic concentrations is facilitated by cyclic-3', 5' adenosine monophosphate. Am Rev Respir Dis 137: 25–30

74. Kidney JC, Dominquez M, Rose M, Aikman S, Chung KF, Barnes PJ (1993) Immune modulation by theophylline: the effect of withdrawal of chronic treatment. Am Rev Respir Dis 147: A772

75. Kidney J, Dominguez M, Taylor P, Rose M, Chung KF, Barnes PJ (1994) Withdrawal of chronic theophylline treatment increases airway lymphocytes. Thorax 49: 396

76. Naclerio RM, Bartenfelder D, Proud D, Togias AG et al (1985) Theophylline reduces the response to nasal challange with antigen. Am J Med 79 [Suppl 6A]: 43–47

77. Naclerio RM, Bartenfelder D, Proud D, Togias AG et al (1986) Theophylline reduces histamine release during pollen-induced rhinitis. J Allergy Clin Immunol 78: 874–876

78. Perrson CGA (1988) Xanthines as anti-inflammatory drugs. J Allergy Clin Immunol 615–616

79. Persson CGA, Erjefält I (1986) Inflammatory leakage of macromolecules from the vascular component into the tracheal lumen. Acta Physiol Scand 126: 615

80. Persson CGA (1986) Role of plasma exudation in asthmatic airways. Lancet 2: 1126

81. Persson CGA, Erjefält I, Anderson P (1986) Leakage of macro-molecules from the guinea pig tracheobronchial microcirculation: effects of allergen, leukotrienes, tachykinins and antiasthma drugs. Acta Physiol Scand 127: 95

82. Erjefält I, Persson CGA (1986) Antiasthma drugs altenuate inflammatory leakage of plasma into airway lumin. Acta Physiol Scand 128: 653

83. Persson CGA, Ekman M, Erjefält I (1979) Vascular anti-permeability effects of betareceptor agonists and theophylline in the lung. Acta Pharmacol Toxicol 44: 216–220

84. Barlinski J, Lockhard A, Frossard N (1992) Modulation by theophylline and enprofylline of the excitatory non-cholinergic transmission in guinea-pig bronchi. Eur Respir J 5: 1201–1205

85. Barnes PJ, Pauwels RA (1994) Theophylline in the management of asthma: time for reappraisal? Eur Respir J 7: 579–591

86. Torvick JA, Borish LC, Beam WR, Kraft M, Wenzel SE, Martin RJ (1994) Does theophylline alter inflammation in nocturnal asthma? Am J Respir Crit Care Med 149: A210

Correspondence: Gilbert E. D'Alonzo, D.O., 925 Parkinson Pavilon, 3401 N. Broad Street, Philadelphia, PA 19140, U.S.A.

Discussion

Moderator: Thank you, Dr. D'Alonzo. I knew that you are good for any surprise. But that bronchial relaxation is paralleled by relaxation of peripheral skeletal musculature was new for me, so people don't throw around their limbs during the night but do so even more during the day. That was very fascinating. Dr. D'Alonzo's paper is open for discussion.

Dr. Poulter: I have two questions. The first one: you very elegantly showed that there were changes in both immunological parameters and physiological ones, comparing 4 pm to 4 am. And indeed you went on to show correlations between the two, implying a cause-and-effect relationship. I just wonder whether you have any evidence as to what that cause-and-effect is. Do changes in the immunology reflect themselves in the physiology or vice versa? The second question is: when these studies comparing 4 pm parameters to 4am parameters were made, were the people measured at 4 pm after lying on their back for 5 hours before those measurements were taken? Because common sense tells me that lying on one's back or one's side or one's front for 5 hours or so may itself have some impact on lung physiology.

Dr. D'Alonzo: Well, it certainly does. And that's been beautifully shown in the literature, that in fact the supine position does alter airway resistance over time. In fact Dr. Morton's study actually tried to control for that by inducing a degree of bronchospasm comparable to what we see at 4 am by a chemical challenge. Dr. Busse's studies were actually performed in the upright position. So these types of factors are being considered and although they haven't been completely dissected out, the data that's been initially presented suggests that body position and the degree of bronchospasm at the time of the challenge has been taken into consideration. Your first question raises an extremely important point: Obviously there could be great gaps between the immunology and the actual functional measurement that has been performed. And we could be missing a great deal of data in between. As far as I know there are no other groups of data that can be compared. And how we can make these associations is the unique challenge for centers that have adequate clinical support but also have strong basic science support and where these two groups can join together to answer these issues and questions. In fact, in the United States there are only a handful of centers, that can accomplish these types of studies.

Dr. Poulter: I think if we understood better the way in which theophylline among other things actually acted, the effect of pharmacological agents on these parameters will in itself give us some information as to which changes are subsequent to others. So we can perhaps dissect the abnormalities occurring in nocturnal asthma by understanding better the way in which these pharmacological agents are working.

Dr. D'Alonzo: Using pharmacology as a way of dissecting out pathologic mechanisms has been used in science for decades if not centuries for this purpose. And theophylline of course is not the only drug with some mechanisms of action totally unknown to us. In fact, the vast majority of drugs have mechanisms of action which remain poorly understood.

Dr. Steinhans: I think after the last 3 presentations it's time to readdress the therapeutic range. Because we heard from Dr. Costello that he's seen an anti-asthmatic effects in the range of 5 to 10 mg/L. The clinicians Dr. Weinberger and Dr. D'Alonzo reported that the control of the airways needs concentrations of about 10 to 15. Could you explain that?

Dr. D'Alonzo: I would like to give my opinion on that. I think that it is extremely important to recognize and think about the data. And certainly, that's just what you said. I personally believe that we should treat with theophylline to clinical efficacy. We should use theophylline in a way in which we see changes in functional parameters, including airflow, but also symptomatic improvement. And in the studies that I've reviewed, you have to do that with plasma theophylline concentrations within the therapeutic range.

Now if we can in addition have certain anti-inflammatory effects, that's wonderful in my opinion. The real question is, is there a subset of patients, in whom we can use theophylline perhaps at a lower level, maybe in combination with other medication, partually to spare the patients from the negative effects of steroids and at the same time enhance the anti-inflammatory actions of these medications. Those types of clinical studies really don't exist at this point and perhaps that will be an interesting pathway of clinical investigation to pursue. At this point in time, I would favour using theophylline clinically in a way in which we can avoid systemic toxicity. So perhaps we should avoid higher therapeutic levels when we are able to, and try to stay within a serum theophylline concentration that clearly demonstrates clinical efficacy. And for many patients that might be a little less than 10, perhaps 7 or 8 mg/L.

Dr. Costello: I'm a clinician, too. I think we're going to identify a very large group of very mild asthmatics who will be effectively treated by low-dose theophylline. I think there's a more severe group with nocturnal asthma, who will need bronchodilating doses of the drug. They will get the additional benefits, one would assume, of the anti-inflammatory activity.

One aspect of the thing we haven't discussed at all here is compliance. It's been mentioned on and off. But it's a very poorly studied subject. And I think one of the major advantages of theophylline is that it is an oral drug. I'm completely contradicting what I would have said 10 years ago. The fact is, particularly children don't like taking inhalers. And one of the few people to have the courage to address compliance is Dr. Cocrane from Guys Hospital who's done a series of studies demonstrating that compliance with inhalers is in general very poor. And I think compliance should improve with an oral non-toxic dose of theophylline in all those mild asthmatics that are out there with their family physicians. The people, the clinicians in this room see the severe asthmatics by and large. There is a huge population of much milder asthmatics out in the population being treated in Britain in general practice and in other ways elsewhere, that may well benefit from long-term low-dose oral theophylline.

Dr. D'Alonzo: So that's a new subset that's worthy of further investigation. I can assure you that with the more moderate to severe asthmatic you have to achieve a serum theophylline level well in the therapeutic range in order to control the disease. In a study that we published looking at Euphylong®, the patient population was defined as having more moderate asthmatic disease. In fact, the median peak flow or FEV1 was about 60% predicted for the group and the variability in airflow approached 30% over each 24-hour time period. So according to the new asthma guidelines, we could classify the group of patients as having more moderate to severe asthma. Yet we were able to control their disease with a modest serum theophylline concentration at night and a much lower level during the daytime without any degree of significant asthma breakthrough. So again perhaps the story might unfold in a way that the more substantial asthmatic has to have a therapeutic level, but a mild group might do very well with a lower than classical therapeutic level. We have to see.

Dr. Lundgren: I just wanted to offer one other explanation why there is discussion about the theophylline range. Perhaps the design of the studies is really quite different. Those who are looking for prevention within one or two days after starting therapy, they may show something, but the long-term studies in large populations may show different things. And maybe you will pick out the effect that Dr. Costello is suggesting if you're following a large group of patients for a long period of time. And I must say, after listening to all these talks this is really what I'm looking for and what I'm waiting to hear about is the longterm studies in large groups of patients which are on corticosteroids, and when put on theophylline, really demonstrate that this is an effective and preventive therapy.

Dr. D'Alonzo: I would like to make one point concerning some information that you brought up. When you use theophylline therapy over a long period of time, I would suggest that there is an impressive reduction in cough. Although it's extremely difficult to measure the degree of mucus production, one way you can indirectly assess this problem is by measuring the degree of coughing. In fact, studies that have assessed cough have clearly shown amelioration with the long term use of theophylline. And getting back to Dr. Steinhans' question. Dr. Costello's group showed with lower doses of theophylline there was a mild improvement of the late asthmatic reaction, and Dr. Pauwel and others have shown near-complete reversal of the late asthmatic reaction. One group used a lower theophylline concentration, the other used a higher one, and perhaps with the higher dose it was interpreted as a mixed anti-inflammatory and bronchodilator effect, but with the lower concentration it is hard to imply a bronchodilator action. Regardless, I'll take both!

Can cytokines in asthma be modulated by theophylline?

A. M. Vignola[1], *M. Spatafora*[2], *G. Chiappara*[1], *A. M. Merendino*[2], *E. Pace*[1], *D. D'Amico*[1], *V. Bellia*[2], and *G. Bonsignore*[1,2]

[1] Istituto di Fisiopatologia Respiratoria, C. N. R. Palermo, Italy
[2] Istituto di Medicina Generale e Pneumologia, Università degli Studi Palermo, Italy

Summary

Theophylline (1,3 dimethylxanthine) is one of the most widely used drugs in the therapy of bronchial asthma and chronic obstructive pulmonary diseases. It is now being increasingly recognized that theophylline is able to down-regulate, both in-vitro and in-vivo, a variety of inflammatory and immune cell functions, such as the late-phase reaction after challenge with the occupational agent toluene diisocyanate and the bronchial responsiveness. The suppressive activity of theophylline is due to its ability to decrease the release of inflammatory mediators and might have important implications with regards to its therapeutic activity in asthma. In this report we summarize the effects of theophylline on TNF-α release, an inflammatory mediator which plays a key role in the pathogenesis of airways inflammation associated with bronchial asthma, by human peripheral blood monocytes (BM) and alveolar macrophages (AM) as well as on the activation and the proliferation of lymphocytes.

The data reported in this paper show that theophylline suppresses TNF-α release and TNF-α gene expression by BM and AM in a dose-dependent fashion at a concentration comparable to the in-vivo therapeutic levels of theophylline in the blood with a maximal inhibition at

2 h and that removal of theophylline after pre-incubation completely abrogates the inhibitory activity on TNF-α release, demonstrating that the continuous presence of the drug is required for suppression of TNF-α release.

In addition, theophylline is also able to affect the function of lymphocytes reducing their activation and proliferation by decreasing the production of IL-2, IL-2 receptor, transferrin and class I MHC antigens.

These data indicate that theophylline is capable of modulating important inflammatory and immune processes and that the therapeutic activity of this drug might be partly related to its down-regulating effects.

Zusammenfassung

Können Cytokine beim Asthma durch Theophyllin moduliert werden? Asthma ist eine chronisch-entzündliche Erkrankung, welche durch die Anhäufung von entzündlichen und immunologischen Effektorzellen (wie Eosinophile, Lymphozyten, Mastzellen und Makrophagen) in den Atemwegen und in der Bronchialschleimhaut charakterisiert ist.

Diese Zellen spielen eine wichtige Rolle bei der Pathogenese der Erkrankung, in dem sie sich gegenseitig beeinflussen, aber auch auf die strukturellen Zellverbände durch Expression von Adhäsionsmolekülen und die Freisetzung von Entzündungscytokinen wirken.

Wir haben angenommen, daß unter diesen Cytokinen der Tumornekrosisfaktor Alpha (TNF Alpha), ein 17-dK Cytokin, das von Lipopolysaccharid (LPS)-stimulierten mononukleären Phagozyten produziert wird, an wichtiger Stelle in die Pathogenese der Erkrankung involviert ist, weil es mit pleomorphen biologischen Aktivitäten ausgestattet ist.

Wir konnten sehen, daß Alveolarmakrophagen (AM) von schweren asthmatischen Patienten spontan große Mengen von TNF Alpha freisetzten (18 ± 3 ng/ml). Weiters sahen wir, daß Endothelin-1 die TNF-Alpha-Freisetzung bei normalen AM steigerte (4 ± 1 ng/ml), bei den AM von schweren Asthmatikern aber die TNF Alpha-Freisetzung aus den AM deutlich verminderte (14 ± 3 ng/ml). Diese Befunde legen nahe, daß beim Bronchialasthma die AM spontan dazu aktiviert werden, TNF Alpha freizusetzen, und daß ihre Fähigkeit vermindert wird, auf heterogene Reize zu reagieren.

Wir haben dann den Effekt von Theophyllin (1,3 Dimethylxanthine) auf die Freisetzung von TNF Alpha aus peripheren Blutmonozyten (BM) und AM untersucht. Diesbezügliche frühere Studien hatten folgendes ergeben:

1. Theophyllin vermindert die Funktion von Entzündungs- und Immunzellen durch Steigerung der intrazellulären Anhäufung von cAMP;
2. Die Freisetzung von TNF Alpha aus mononuklearen Phagozyten wird durch den cAMP Spiegel reguliert.

Theophyllin reduziert dosis-abhängig den bioaktiven TNF-Alpha-Ausstoß aus menschlichen BM. Eine signifikante Hemmung wurde bei 100 µM beobachtet (40,8 ± 5,9% bei Kontrollen; p < 0,01) und bei 50 µM (59,2 ± 4,8% bei Kontrollen; p < 0,05), während die Aktivität von Theophyllin bei 10 µM (71,2 ± 8,9% bei Kontrollen) nicht statistisch signifikant war. Die Aktivität erreichte ihr Maximum nach 2 Stunden und fiel nach 4 Stunden (59,0 ± 5,2% der Kontrollen) und nach 24 Stunden (89,1 ± 3,1% der Kontrollen) wieder ab. Eine Northern Blot-Analyse von RNA, die aus menschlichen BM extrahiert worden waren, zeigte, daß Theophyllin imstande war, die TNF Alpha-Gen-Expression zu vermindern. Zumal TNF Alpha an wichtiger Stelle in der Pathogenese der bronchialen Hyperreaktivität und des Asthma involviert ist, lassen diese Befunde vermuten, daß die therapeutische Aktivität von Theophyllin zumindest teilweise über seinen Effekt auf den TNF Alpha Ausstoß von mononukleären Phagozyten erklärbar ist.

Introduction

Bronchial asthma is a chronic inflammatory disease characterized by the increased responsiveness of the airways to a variety of stimuli. Current concepts on the pathogenesis of this disease suggest that inflammatory and immune effector cells (eosinophils, lymphocytes, mast-cells and macrophages) are activated in the airways and within the bronchial mucosa and release heterogeneous mediators of inflammation (cytokines, arachidonic acid metabolites) and injury (ECP, free oxygen radicals).

Theophylline (1,3 dimethylxanthine) is one of the most widely used drugs in the therapy of bronchial asthma and chronic obstructive pulmonary diseases. The therapeutic activity of theophylline has been classically related to its bronchodilator effect [1] and to its ability to improve diaphragmatic contractility [2]. However, it is now being increasingly recognized that theophylline down-regulates, both in-vitro and in-vivo, a variety of inflammatory and immune cell functions, probably by virtue of its ability to increase the intracellular concentration of cyclic AMP (cAMP), a second mediator endowed with potent suppressor activity on inflammation. Hence, it is conceivable that the therapeutic activity of theophylline is related, at least partly, to its effects on inflammation. This hypothesis is supported by the observation that theophylline has a more pronounced protective effect on the late asthmatic reaction after allergen challenge than on the immediate reaction [3] and that, although theophylline does not modify bronchial hyperreactivity, it is able to inhibit the late phase reaction after challenge with the occupational agent toluene diisocyanate (TDI) [4]. Moreover, in a more recent study, oral slow-release theophylline was able to inhibit the allergen-induced asthmatic reaction and the associated increase of bronchial responsiveness [5]. Thus, the suppressive activity of theophylline on the release of inflammatory mediators might have important implications with regards to its therapeutic activity.

The aim of this report will be to summarize the current knowledge on the effects of theophylline on mononuclear phagocytes and lymphocytes; the effects on the other inflammatory and immune cells involved in the pathogenesis of bronchial asthma will be covered in other contributions published in this same issue.

Effects of theophylline on mononuclear phagocytes

Mononuclear phagocytes play an important regulatory role in the development of immune and inflammatory reactions associated with asthma [6, 7]. In asthma, monocytes and airways macrophages (AM) are increased in numbers in the bronchial mucosa [8] and are in an activated state, as demonstrated by their low cellular density [9] and by their ability to release lysosomal enzymes, arachidonic acid metabolites, platelet-activating factor, cytokines and oxygen free radicals [12] after in vivo [10] or in vitro [11] challenge with specific allergens or nonspecific stimuli. This functional activation is associated with phenotypical changes such as an increased expression of adhesion molecules [13] and the induction of the expression of the low affinity receptor for IgE (FceRIIb/CD23b) [14].

The state of permanent functional activation of AM during the inflammatory process associated with bronchial asthma may also lead to the reduction of the in-vitro releasability of these cells. Consistent with this hypothesis is the evidence that, while endothelin-1 (a mitogenic and bronchoconstrictor mediator involved in the pathogenesis of asthma) and bacterial lipopolysaccharide (LPS) increase the expression of cell surface markers and the release of tumour necrosis factor-a (TNF-α) and fibronectin by alveolar macrophages from normal subjects, they decrease the release of the same mediators by AM recovered from severe asthmatic patients [15]. In addition to the reduced response to up-regulating stimuli, the constant activation of AM in asthma causes a lower sensitivity of these cells to physiological suppressive mediators such as interleukin-4 (IL-4), a cytokine capable of decreasing the production of other inflammatory cytokines including IL-1, TNF-α, and IL-6 [16] and the release of oxygen free radicals. Hence, the regulation of the production of cytokines and other mediators released by AM may play a crucial role in the activity of drugs used in the therapy of bronchial asthma.

Among the mediators released by mononuclear phagocytes located in the airways, TNF-α, an important inflammatory cytokine produced by LPS-stimulated cells [17–20] likely plays a major role in the pathogenesis of bronchial inflammation and hyperresponsiveness. In addition to its own inflammatory activities, TNF-α induces the release of other mediators, such as the powerful neutrophil chemotactic and activating factor, interleukin 8 [21]. Moreover, TNF-α increases neutrophil adherence to endothelial cells by increasing the expression of adhesion molecules on the surfaces of both interacting cells [22] and promotes the transendothelial passage of neutrophils [23]. Not only TNF-α may be importantly involved in the development of non-specific bronchial inflammation and hyperreactivity, but also it may participate to the lung inflammatory processes that follow the inhalation of a specific antigen in allergic asthmatic patients. TNF-α is released by AM of allergic asthmatic patients after the development of late asthmatic reaction induced by allergen inhalation [11]. Moreover, the IgE receptor triggering by specific antigens or anti-IgE antibodies induces, in addition to the release of histamine and arachidonic acid metabolites, the expression of the TNF-α gene and the release of bioactive TNF-α by rat basophils and rat lung explants [24]. Since the release of TNF-α by mononuclear phagocytes is regulated by the intracellular levels of cAMP, we investigated the effects of theophylline on TNF-α release by human peripheral blood monocytes (BM) and alveolar macrophages (AM). We found that theophylline suppresses TNF-α release and TNF-α gene expression in a dose-dependent fashion and at concentrations comparable to the invivo therapeutic levels of theophylline in the blood [25] (Fig. 1). In addition, we found that the inhibition of TNF-α release by theophylline is maximal at 2 h, compared to subsequent time-points (4 and 24 h) (Fig. 2) and that removal of theophylline after pre-incubation completely abrogates the inhibitory activity on TNF-α release, demonstrating that the continuous presence of the drug is required for suppression of TNF-α release. Finally, we did not detect any difference on the susceptibility of AM and BM to the inhibitory activity of theophylline on TNF-α release, indicating that differentiation processes of mononuclear phagocytes do not interfere with the regulatory activity of theophylline (Fig. 3).

The activity of theophilline on the bactericidal functions of alveolar macrophages have also been investigated: the administration of oral theophylline for 14 days was shown to determine an impairment of this

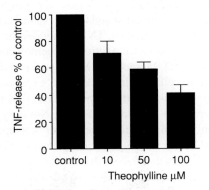

Fig. 1. Effect of theophylline on TNF-α release by LPS-stimulated human blood monocytes. After purification procedures, the cells were stimulated with LPS (1 μg/ml) in the absence or presence of theophylline at serial concentration. Data represent mean ± SEM of five separate experiments

function due to the reduction of the release of hydrogen peroxide [26]. While it is conceivable that this activity of theophylline affects host defence against bacterial infection within the lungs, it may also prevent the potential toxic effects of reactive oxygen species on pulmonary structural and inflammatory cells.

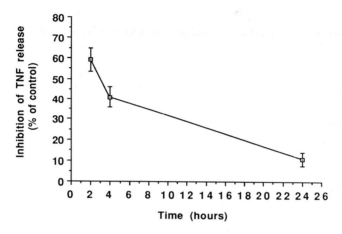

Fig. 2. Kinetic analysis of theophilline-induced inhibition of TNF-α release by human blood monocytes. Data represent mean ± SEM of five separate experiments

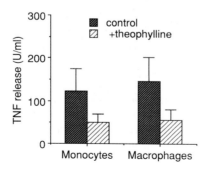

Fig. 3. Effect of theophylline (2 hr; 100 μM) on LPS-induced release of TNF-α by rat blood monocytes and alveolar macrophages

Effects of theophylline on lymphocytes

In the last years, several investigators have evaluated the ability of theophylline to modulate the activity of peripheral lymphocytes. In this regard, theophylline activates T suppressor systems in vivo, as demonstrated by the reduction of allograft reaction in humans and experimental animal models [27]. In addition, Pardi et al. have shown that the infusion of aminophylline at the usual therapeutic dosage significantly reduces the proliferation of lymphocytes after in vitro challenge with lectins and other mitogens; interestingly, this effect was associated with an increase of cAMP levels in circulating lymphocytes [28]. Consistent with these findings, Scordamaglia et al. demonstrated that theophylline decreases the proliferation of T-cells and T-cell clones after stimulation with anti-CD3 monoclonal antibody or phytohemagglutinin (PHA) by reducing the production of the powerful T-cell growth factor, IL-2 [28a]. The inhibitory activity of theophylline on lymphocyte proliferation may be related to its ability to decrease, in a dose-dependent fashion, the PHA-induced expression of surface molecules (such as the IL-2 receptor, transferrin and class I MHC antigens) importantly involved in the process of T-cell activation and proliferation [29].

In addition to its effects on T-cells, theophylline has also been shown to inhibit the spontaneous cytotoxic activity of large granular lymphocytes (LGLs) against tumor cells and virus-infected cells (natural killer activity). This activity is also due to the increase of the intracellular cAMP levels [30, 31]. It is currently unknown whether this effect is due to phenotypical modifications of LGLs with a relative increase of LGLs with low lytic activity, to reduced binding of LGLs to their targets, or to reduced release of cytotoxic mediators.

Conclusions

In conclusion, theophylline is a powerful molecule capable of inhibiting TNF-α release and TNF-α gene expression by mononuclear phagocytes and to modulate the function of immune-effector cells. Further studies are in progress to evaluate the in-vivo and in-vitro activity of the drug on this and other functions of inflammatory cells as well as on its ability to inhibit the production of other inflammatory cytokines involved in the pathogenesis of bronchial asthma.

References

1. Piafsky KM, Ogilvie R (1975) Dosage of theophylline in bronchial asthma. N Engl J Med 292: 1218–1222

2. Aubier M, De Troyer A, Sampson M, Macklem PT, Roussos C (1981) Aminophylline improves diaphragmatic contractility. N Engl J Med 305: 249–252

3. Pauwels RA (1987) The effects of theophylline on airway inflammation. Chest 92: 32s–37s

4. Mapp CE, Boschetto P, Dal Vecchio L, Crescioli S, De Marzo N, Paleari D, Fabbri LM (1987) Protective effect of antiasthma drugs on late asthmatic reactions and increased airway responsiveness induced by toluene diisocyanate in sensitized subjects. Am Rev Respir Dis 136: 1403–1407

5. Crescioli S, Spinazzi A, Plebani M, Pozzani M, Mapp CE, Boschetto P, Fabbri LM (1991) Theophylline inhibits early and late reactions induced by allergens in asthmatic subjects. Ann Allergy 66: 245–251

6. Metzger WJ, Zavala D, Richerson HB, Moseley P, Iwamota P, Monick M, Sjoerdsama K, Hunninghake G (1987) Local allergen challenge and bronchoalveolar lavage of allergic asthmatic lungs. Am Rev Respir Dis 135: 433–440

7. Joseph M, Tonnel AB, Torpier G, Capron A, Arnoux B, Benveniste J (1983) The involvement of IgE in the secretory processes of alveolar macrophages from asthmatics patients. J Clin Invest 71: 221–226

8. Poston R, Chanez P, Lacoste JY, Litchfield P, Lee TK, Bousquet J (1992) Immunohistochemical characterization of the cellular infiltration of asthmatic bronchi. Am Rev Respir Dis 145: 918–921

9. Chanez P, Bousquet J, Couret I et al (1991) Increased numbers of hypodense alveolar macrophages in patients with bronchial asthma. Am Rev Respir Dis 144: 923–930

10. Arnoux B, Joseph M, Simoes MH, Tonnel AB, Duroux P, Capron A, Benveniste J (1987) Antigenic release of PAF-acether and β–glucuronidase from alveolar macrophages of asthmatics. Bull Eur Physiopathol Respir 23: 119–124

11. Gosset P, Tsicopoulos A, Wallaert B, Vannimenus C, Joseph M, Tonnel AB, Capron A (1991) Increased secretion of tumor necrosis factor alpha and interleukin-6 by alveolar macrophages consecutive to the development of the late asthmatic reaction. J Allergy Clin Immunol 88: 561–571

12. Calhoun WJ, Reed HE, Moest DR, Stevens CA (1992) Enhanced superoxide production by alveolar macrophages and air-space cells, airway inflammation, and alveolar macrophage density changes after segmental antigen bronchoprovocation in allergic subjects. Am Rev Respir Dis 145: 317–325

13. Chanez P, Vignola AM, Lacoste P, Michel FB, Godard P, Bousquet J (1993) Increased expression of ICAM-1 and LFA-1 on alveolar macrophages from asthmatics. Allergy (in press) 48: 576–580

14. Gosset P, Tsicopoulos A, Benoit W, Joseph M, Capron A, Tonnel AB (1992)

Tumor necrosis alpha and interleukin-6 production by human mononuclear phagocytes from allergic asthmatics after Ig-E dependent stimulation. Am Rev Respir Dis 146: 768–784

15. Chanez P, Vignola AM, Springal DR, Campbell AM, Farce M, Polak J, Michel FB, Godard FB, Bousquet J (1993) Endothelin and airway cells in asthma. Am Rev Respir Dis 147: A517

16. Yanagawa H, Sone S, Sugihara K, Tanaka K, Ogura T (1991) Interleukin-4 down-regulates Interleukin-6 production by human alveolar at protein and mRNA levels. Microbiol Immunol 35: 879–893

17. Carswell EA, Old LJ, Green S, Fiore N, Williamson B (1975) An endotoxin-induced serum factor that causes necrosis of tumors. Proc Natl Acad Sci USA 72: 3666–3670

18. Beutler B, Cerami A (1987) Cachectin: more than a tumor necrosis factor. N Engl J Med 316: 379–385

19. Lynch JP III, Toews GB (1989) Tumor necrosis factor-a. A multifaceted mediator of inflammation. Chest 96: 457–459

20. Taffet SM, Singhel KJ, Overholtzer JF, Shurtleff SA (1989) Regulation of tumor necrosis factor expression in a macrophage-like cell line by lipopolysaccharide and cyclic AMP Cell Immunol 120: 291–300

21. Matsushima K, Morishita K, Yoshimura T, Lavu S, Kobayashi Y, Lew W, Appella E, Kung HF, Leonard EJ, Oppenheim JJ (1988) Molecular cloning of a human monocyte-derived neutrophil chemotactic factor (MDNCF) and the induction of MDNCF mRNA by interleukin 1 and tumor necrosis factor. J Exp Med 167: 1883–1893

22. Salyer JL, Bohnsack JF, Knape WA, Shigeoka AO, Ashwood ER, Hill HR (1990) Mechanisms of tumor necrosis alpha alteration of PMN adhesion and migration. Am J Pathol 136: 831–841

23. Moser R, Scleiffenbaum B, Groschurth P, Fehr J (1989) Interleukin 1 and tumor necrosis factor stimulate human vascular endothelial cells to promote transendothelial neutrophil passage. J Clin Invest 83: 444–455

24. Ohno I, Tanno Y, Yamauchi K, Takishima T (1990) Gene expression and production of tumor necrosis factor by a rat basophilic leukemia cell line (RBL-2H3) with IgE receptor triggering. Immunology 70: 88–93

25. Rall TW (1980) The Xanthines. In: Gilman AG, Goodman LS, Rall TW, Murad F (eds) The pharmacological basis of therapeutics. Mac Millan, New York, pp 592–607

26. O'Neill SH, Sitar DS, Klass DJ, Taraska VA, Kepron W, Mitenko PA (1986) The pulmonary disposition of theophylline and its influence on human alveolar macrophage bactericidal function. Am Rev Respir Dis 134: 1225–1228

27. Shapira Z, Shoat B, Boner G, Levi J, Joshua H, Servadio C (1982) Theophylline: a possible immunoregulator of T-cells. Transpl Proc 14: 113–118

28. Pardi R, Zocchi MR, Ferrero E, Ciboddo GF, Invernardi L, Rugarli C (1984) In vivo effects of a single infusion of theophylline on human peripheral blood lymphocytes. Clin Exp Immunol 57: 722–728